Charles H[enry] Stowell

The Essentials of Health

A Text Book of Anatomy, Physiology, Hygiene, Alcohol, and Narcotics

Charles H[enry] Stowell

The Essentials of Health
A Text Book of Anatomy, Physiology, Hygiene, Alcohol, and Narcotics

ISBN/EAN: 9783744670012

Printed in Europe, USA, Canada, Australia, Japan

Cover: Foto ©berggeist007 / pixelio.de

More available books at **www.hansebooks.com**

THE

ESSENTIALS OF HEALTH

A Text-Book

ON

ANATOMY. PHYSIOLOGY, HYGIENE, ALCOHOL, AND NARCOTICS

BY

CHARLES H. STOWELL, M.D.

LATE PROFESSOR OF HISTOLOGY AND MICROSCOPY, AND ASSISTANT PROFESSOR OF
PHYSIOLOGY, UNIVERSITY OF MICHIGAN.
AUTHOR OF "STUDENT'S MANUAL OF HISTOLOGY," "MICROSCOPICAL DIAGNOSIS." "THE
STRUCTURE OF TEETH," "A HEALTHY BODY," "A PRIMER OF HEALTH," ETC.
LATE EDITOR OF "THE MICROSCOPE"
EDITOR OF "THE NATIONAL MEDICAL REVIEW."

FULLY ILLUSTRATED

WITH ORIGINAL SKETCHES BY THE AUTHOR

SILVER, BURDETT & COMPANY

NEW YORK .   BOSTON   CHICAGO

1892

PREFACE.

THIS book has been prepared for pupils who have studied the preceding number of this series, A HEALTHY BODY, or a work of like grade. It will meet the requirements for advanced study in public and private schools, in high schools and academies, and is well adapted to the needs of the general reader.

The illustrations are from original pen sketches drawn by the author. In nearly every instance these sketches were made from specimens especially prepared by the author for this series of books.

The treatment of the subject of anatomy is sufficient to form the foundation for physiological study; physiology is given with sufficient fullness to elucidate the principal features of the subject, and to enable the pupil to comprehend easily the laws for the preservation of health; hygiene is dwelt upon with such completeness that the ordinary laws of health need not be ignorantly broken; while the subjects of alcohol and tobacco are discussed with such care and in such detail as to show that the use of these narcotics by the young may not only diminish their powers and dwarf them physically. but may even altogether arrest the mental, moral, and physical development. No thoughtful person can study the effects of alcohol and tobacco on the growing body without becoming seriously impressed with the great power for evil which these drugs possess. The evidence is already convincing, and is daily accumulating, that the future welfare of our youth, and, therefore, the future welfare of our country, demands that both these poisons be placed beyond the reach of the young.

A knowledge of the fundamental principles of anatomy and physiology, together with a knowledge of those things which are desirable and of those which should be shunned, will do much toward the development of a noble manhood and womanhood.

CHARLES H. STOWELL.

WASHINGTON, D. C.,
July, 1892.

CONTENTS.

6 CONTENTS.

INTRODUCTION.

THE anatomist teaches that the body is composed of *tissues* and *fluids*. A number of tissues are often so united that the part can best perform some special work or function. Such a part is called an *organ ;* as, the eye is the organ of sight, and the ear, the organ of hearing. Tissues which have a similar structure are grouped together into *systems*. Thus we have the muscular system, and the nervous system. Tissues are also grouped according to their functions, each group receiving the name of an *apparatus*. To illustrate, we have the digestive apparatus, and the respiratory apparatus.

The histologist teaches that each tissue is composed of most minute parts. These are in the form of *fibrils* and *cells*. To study these anatomical elements requires the aid of the highest powers of the microscope. Yet a study of the life history of a single cell often reveals many of the wonderful phenomena of the whole body. The microscope brings man in closest relations with nature, and renders clear many of her mysterious processes.

The chemist teaches that the body is composed of a number of *chemical elements*. These include such well-

known elements as carbon, hydrogen, oxygen, and nitrogen, together with a number of others. These elements
form certain combinations and give rise to certain
chemical processes. The chemist also informs us that •
the arrangement of these elements can be disturbed by
certain forces and a new arrangement formed, by which
entirely new substances appear. Thus we find that the
growing plant can take carbonic acid gas (which is composed of carbon and oxygen) and water (which is
composed of hydrogen and oxygen), break up the arrangement of their elements, and rearrange them so
that starch (which is composed of carbon, oxygen, and
hydrogen) is deposited in the leaves. Starch is ap-
parently unlike the two elements from which it is
formed; yet it is composed of the same chemical elements, only these are differently arranged and in different proportions. The proportions are different
because when the growing plant breaks up the carbonic acid gas it sets the oxygen free, and unites the
carbon with the hydrogen and oxygen of the water.
By adding more hydrogen and oxygen to the starch, the
living plant can change the starch to sugar.

The physiologist teaches of the active phenomena pre-
sented by living beings. The peculiar action of a particular organ is called its *function*. It is evident that
physiology is best studied by carefully observing living
creatures, — by the direct observation of nature. From
the study of a dead tissue we are unable to infer, by any
process of reasoning, what its living actions were. For
instance, there is nothing about the structure of a nerve

from which we can reason that it conveys a sensation of pain, or a stimulus for the contraction of a muscle. Neither is there anything in the structure of the gastric glands from which we might reason that they secrete a digestive fluid. Physiological properties are not ascertained by anatomical examination.

The hygienist teaches a system of principles and rules, based upon a knowledge of anatomy, physiology, and chemistry. These rules have for their main object the preservation of the health; it may be the health of the individual, or that of an entire community. Hygiene is the practical result obtained from a study of the science of medicine. By a proper observance of its laws, the individual is enabled to accomplish his best work; while he is also able to prevent much sickness and to reduce its severity when present.

The laws of health should inform us not only concerning those things which tend directly to the promotion of health, but also concerning those which must be avoided if health is to be maintained. As the evidence is overwhelming that alcohol and tobacco exert a most pernicious influence on the growing body, the effects of these drugs should be carefully and fully set forth, that the individual may be informed of the danger which attends their use. These drugs are specifically named because it is believed that the very wide use they have gained is largely due to ignorance of their nature. What is said of alcohol and tobacco may with propriety be said of several other narcotics, as opium, the use of which is not so prevalent in this country. All these stimulants and narcotics tend to sap the vital forces, weaken the will, and undermine the moral character.

The great value of sound health, of the individual and of the community, has given rise to organized effort in the form of Boards of Health in all our large cities, and in most if not all of the States of the Union. These Boards and Sanitary Commissions have done much, not only to arrest the spread of disease but to prevent its appearance. The subject is one of such vital importance that it appeals to intelligence everywhere. It has enlisted the best thought of the wisest men and women for many years. Manifestly, therefore, it is unnecessary to offer any apology for the prominence given to the essentials of health in the pages of this volume.

ESSENTIALS OF HEALTH.

CHAPTER I.

CELLS.

General Description. The whole body is composed largely of cells; while each individual part consists of cells quite characteristic in shape and size. By studying these cells we learn much of the anatomy and physiology of the whole system.

Some cells are so minute that very high powers of the microscope are required to see them, while others are nearly large enough to be seen with the unaided eye. In shape, there is the greatest variation. There are spherical, oval, and spindle-shaped cells; cells with branches extending in various directions; and still other cells with six equal sides. In color, there are the extremes from the black to the colorless; and from the brown to the yellowish green. There exists, therefore, a great variety in the shape, size, and color of cells.

Their Structure. Living cells consist of a transparent, jelly-like material, called protoplasm. The microscope shows that there are two parts to a cell: the body, or the

greater part of the cell ; and the nucleus, or the smaller
part in the centre. The nucleus is usually spherical
or oval, and, with few exceptions, is found in all cells.
In rapidly growing cells two or more nuclei are often
found. The nucleus of a cell can be shown very clearly
by adding a carefully prepared solution of carmine, and
then examining with the microscope. The carmine
stains each nucleus bright red, but does not affect the
body of the cell.

The accompanying colored plate illustrates the results
of carmine-staining on a variety of cells : (1) mucous
cells, found wherever mucus is secreted; (2) columnar
cells, surrounding the villi of the small intestine ;
(3) cells from connective tissue; (4) ciliated cells from
the trachea; (5) flattened epithelium from the mucous
membrane of the mouth ; (6) liver cells ; (7) cells
from the surface of the body ; these cells have no
nuclei ; (8) cells from the salivary glands ; (9) cells
lining the cavities of the heart ; (10) pigment cells from
the eye.

The Life of a Cell. It is probable that the great majority of cells are, comparatively speaking, short-lived.
We must remember that the body is constantly and
rapidly changing. Each movement of the body, each
activity of a part, must cause a wear and waste of
tissue ; and this loss must be replaced by new material within a short time.

There are many ways of showing that the body is ever
wasting away. If a drop of saliva be placed under the
microscope, a vast number of thin cells can be seen.
These cells come from the mucous membrane lining the
mouth. The motion of the tongue, lips, and cheeks, as

PLATE I.

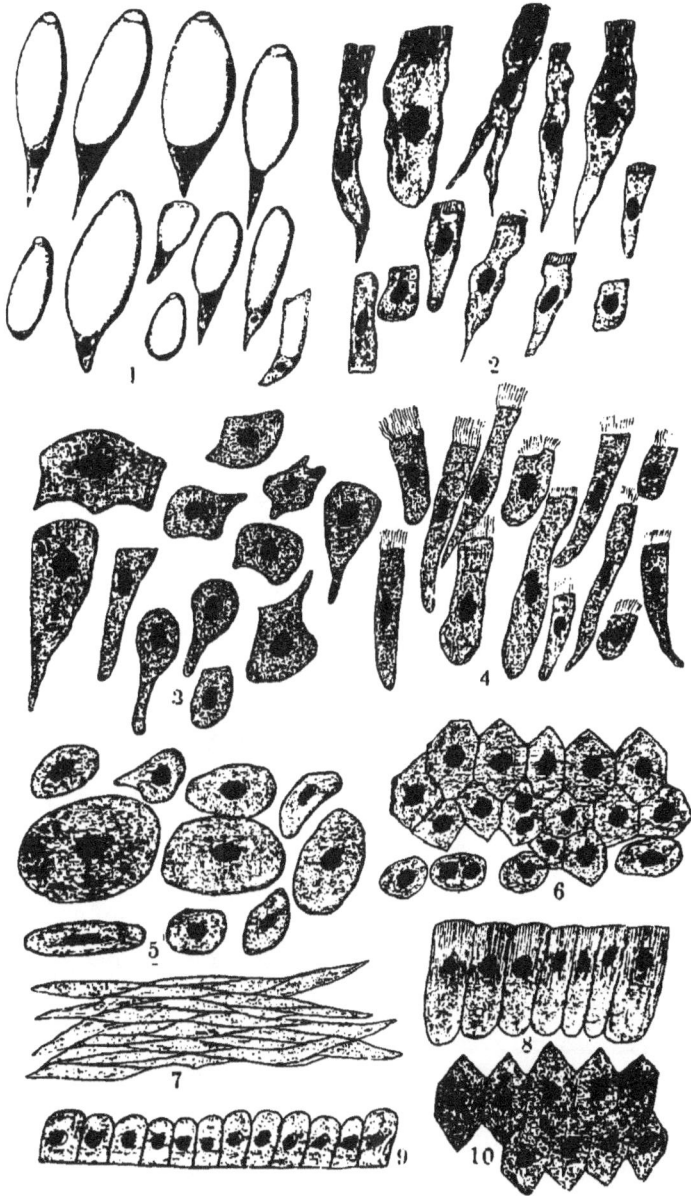

in speaking, eating, and drinking, remove vast numbers
of these bodies. Then again, the surface of the whole
body is covered with cells, many layers deep. The
outer cells are easily removed, by the friction of the
clothing, and by the use of the sponge and towel at
the daily bath. In this way immense numbers of cells
are being constantly destroyed, while new ones are as
rapidly being formed beneath the surface to take their
places.

A more familiar example will illustrate this point.
The finger nails are composed of cells so minute that
a high power of the microscope is required to see them.
Each paring of the nail, therefore, must remove vast
numbers of these cells; and yet, how rapidly even this
hard structure grows. Thus we learn that the body is
ever changing; the old, worn-out, and useless material
being constantly cast off, and the new as regularly taking its place.

Some cells are much longer-lived than others. It is
probable that the cells found in such hard tissues as
bone and cartilage undergo comparatively slow changes,
while the cells in some of the glands change with great
rapidity. In fact, the whole life history of a cell in
some of the most active glands may be covered by
a few hours.

Growth and Development. Cells increase both in size
and number. After reaching certain dimensions, however, they cease to grow. They may then either maintain
that size for the remainder of their life, or they may,
by a peculiar process of division, become temporarily
smaller. This process is called cell division. When a
cell is about to divide, its nucleus becomes constricted in

the center, assuming a dumb-bell shape. This con-
striction increases until the nucleus becomes divided
into two nuclei. Then the body of the cell undergoes the same change in form until it has divided

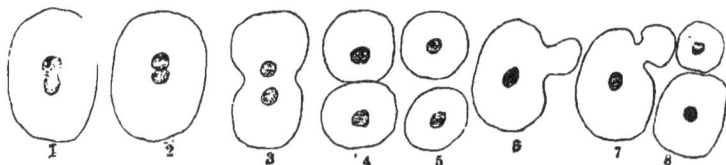

FIG. 1.   Diagram illustrating the division of cells : (1, 2, 3, 4, 5,) by
cell division ; (6, 7, 8,) by budding.

into two cells, with one nucleus for each cell. Another
method of division is that where a process, or bud,
protrudes from the body of the cell. Soon this sepa-
rates from the original cell, and a nucleus is developed
within it.

Some Cells have Motion. The great majority of the
cells in the body are fixed and cannot alter their shape
or position. There are some, however, that not only
have the power to change their shape, but also to move
from place to place. These movements are known as
the " amœboid movements," so named from an animal
called the amœba.

The amœba is the lowest form of animal life. It
is of jelly-like consistence, and averages from the $\frac{1}{500}$
to the $\frac{1}{2500}$ of an inch in diameter. It is found in
stagnant water, and in water in which there is decaying animal matter. The amœba is an object of intense
interest to all physiologists, because it represents both
a single cell and a whole individual. It is remarkable for its constant and rapid changes of form,

causing it to move about in any direction. The amœba
is an animal, the lowest in the scale; yet it moves;

FIG. 2. Various forms assumed by an amœba. These sketches were
made from the same amœba, at intervals of a few seconds.

it takes nourishment; it reproduces its own kind;
and it dies.

The Function of Cells. Certain cells are set apart
to perform certain work, and they can do no other.
These groupings are called glands, tissues, and organs.
The cells of the salivary glands can take digested food
from the blood, and change it into the tissue of the salivary glands. The cells of muscle can take something
from the blood and build from it true muscular tissue.
The cells of the skin take the same nourishment from
the blood, and make from it the soft covering for the
body. Thus, while a person may eat only one kind of
food, it is possible for this food to be changed into all
the various structures of the body.

This is not the work of any one organ; it is the
combined work of the cells of all the organs and tissues of the body. Each cell, therefore, can take material from the blood, and change it into its own
structure. But the cell can do even more than this;
it can take material from the blood and change it into
a substance unlike its own. For instance, a cell in
one of the salivary glands can take some material

brought to it by the blood, and completely change it into saliva; or a cell in one of the glands of the stomach can take material from the blood and change it into gastric juice. We learn that each cell in the body has its own work to do, and that this work is constant and rapid.

ALCOHOL AND TOBACCO.

It must be evident that deficiency of food, or any injurious substance brought into contact with the cells, must seriously interfere with the health of the body. Any substance which will contribute to the normal growth and activity of the cells is greatly to be desired, while anything which would impair these should be avoided. We should remember, also, that all young and rapidly growing cells are much more sensitive to foreign substances than are those which are fully developed. Alcohol and nicotine, being foreign substances and poisons, retard the healthy action and growth of the cells.

QUESTIONS.

1. What is said about the size, shape, and color of cells?
2. Give the structure of cells.
3. Give illustrations proving that the body is ever wasting away.
4. How do cells increase in number?
5. Give a description of the amœba.
6. What is the work of the cells?
7. What can each individual cell do?
8. How do alcohol and tobacco affect the growth of the cells?

CHAPTER II.

FOODS.

Waste and Repair. We must keep in mind that each activity of the body is followed by a waste of material, and that this change is constant throughout life. As the body is always wasting away, so should it always be undergoing repair. The processes of waste and repair do not always bear the same relation to each other.

Early in life the building up greatly exceeds the breaking down; more material is supplied than is worn out and removed; the processes of repair exceed those of waste, and the body grows and develops. Later in life the repair and waste are nearly balanced, and for a number of years the form and weight remain about the same. As old age comes on, the weight diminishes and all the forces of man become less active. The waste now exceeds the repair.

From the food we eat, the body must obtain the materials for building its structure and for keeping it in repair. It is clear, viewed in this light, that the subject of foods is of the utmost importance.

Uses of Foods. The first use of foods, therefore, is to supply material from which the body may be built up and kept in repair. Other uses are, to supply heat for the body; to give strength ; and to dissolve substances and aid in their distribution. If it is the nature of a substance to injure a healthy organ, or in any way interfere with its action, it cannot be called a food.

Classification of Foods. All three of Nature's kingdoms are called upon to furnish articles of food. The principal articles obtained from the mineral kingdom are water and salt; from the vegetable kingdom, such cereals as wheat, corn, and oats, and a large number of vegetables, such as potatoes and fruits ; from the animal kingdom, the various meats, milk, and eggs.

For purposes of study, foods are divided into the organic and inorganic. The organic foods are obtained from living substances, or from things which once had life. The inorganic foods are derived from such inorganic substances as the air, the earth, and the water.

THE INORGANIC FOODS.

The two principal inorganic foods are water and salt. These are used by themselves, or are added to the food in cooking.

Salt. Salt is found in all the tissues and organs of the body, except the enamel of the teeth. It is estimated that there is nearly one quarter of a pound of salt in the entire body. In a small amount, salt is present in nearly all the organic foods in use, but not in sufficient quantity to meet the demands of the system. That salt is a necessary food is indicated by the natural craving for it, not only in man, but in the lower animals as well. Animals on the prairies, like the buffalo for instance, will travel many miles in search of salt; while the sheep come quickly to the farmer's call, expecting some of this necessary food. These animals, together with others living upon the grasses, do not obtain a sufficient supply of salt with their food; while those animals living principally upon meats receive a proper amount, as it is already in the meat itself. Such animals may even have a repugnance for salted meats.

Salt gives a flavor to the food. Food may be very nutritious, yet if it be tasteless it is not eaten readily, and is digested with difficulty. Salt stimulates the appetite, excites a flow of the saliva and gastric juice, and thus aids in the whole process of digestion.

Water. Water constitutes nearly three fourths of the weight of the entire body. It is universally present in all the tissues and fluids of the body. There are many reasons why water is so important. All the food must be dissolved before the processes of digestion and absorption can be completed. The water in the tissues holds in solution numerous substances, which represent some of the waste materials of the body.

Water gives elasticity to the bones, the muscles, the tendons. and the other tissues. Through the blood and tissues it becomes a circulating medium for conveying the foods, which are held in solution, to all parts of the body, and for taking away from the tissues the worn-out and useless ingredients. It is due to the presence of water that each tissue has its special consistency. Water is the most important substance used as food, as it is the one universal solvent. It thus becomes the medium for carrying dissolved materials from place to place, leaving and taking, as nature directs. The craving for water is greater than for any other food, and a person will die sooner, if deprived of it, than if deprived of solid food.

A large quantity of water is taken into the system during each day. Some of this is taken purposely, as a drink, while a large amount is taken with the food. To prove the truth of this latter statement, we have only to recall the fact that one half the weight of beef, three fourths the weight of potatoes, and nine tenths the weight of milk, consist of water. A healthy man takes, on an average, about two quarts of water each day.

Sources of Water. Rain water most closely resembles distilled, or chemically pure water; it usually contains a small amount of carbonic acid gas. Spring water contains a considerable amount of mineral substances and carbonic acid gas, the latter giving to spring water its fresh taste. The carbonic acid aids in dissolving the mineral substances as the water permeates the soil. Spring water contains but little oxygen; therefore vegetable organisms usually flourish in it, while animal life, which requires much oxygen, is poorly represented. Spring water may bubble to the surface of the earth, or some mechanical device, as a pump, may be necessary to bring it within reach.

The running water of rivers does not contain so much mineral matter as does spring water. When brought to the surface, spring water rapidly gives off its carbonic acid, causing a deposit of some of the mineral substances. Running water absorbs oxygen readily from the air, and being deficient in carbonic acid and rich in oxygen, it affords the necessary conditions for animal life. Drinking water is obtained chiefly from springs, and usually some form of a pump is necessary to bring it to the surface.

The Purity of Drinking Water. Drinking water should be colorless and without the slightest odor. Chemically pure water, however, is not pleasant to the taste, it lacks the snap and tartness of spring water. Then, too, the presence of some minerals in solution is useful to the system. Lime is important in the formation of teeth and bone, and when drinking water does not contain lime in excess, then it must be regarded as beneficial, especially so in early life, when the tissues are developing.

On the other hand, lead is a very dangerous ingredient of water. Water that has stood in lead pipes should never be used for drinking purposes. If it is necessary to use lead pipes, then the water should be kept running, or a large quantity drawn off before any is taken for use. There is no reason for supposing, simply because water has no odor and looks clear, that it does not contain in solution substances of a most poisonous nature.

Organic matter may be present in a state of decomposition. When there is any danger of this, it is much better to boil the water for a long time, and thereby destroy the minute germs. During an epidemic of typhoid fever it is a wise precaution to drink no water that has not been so treated. When cool, after boiling, a little lime juice may be added to the water to correct the insipid taste, as the boiling drives off the carbonic acid gas. Do not use water that may be near some source of filth, and be sure that the well or spring is located at a safe distance from any decaying organic matter.

While the system daily demands a large amount of water, yet too much is injurious, especially if taken with the meals. When the body is in a healthy condition, the demand for water is easily satisfied. Remember it is best to drink sparingly at mealtime, and more freely between meals. It is very unwise to use ice-cold water at any time, and it is especially bad to take it with a meal.

There are other inorganic foods besides water and salt, — as the various chemical salts, such as the salts of soda, potassa, and lime.

QUESTIONS.

1. What relation do waste and repair bear to each other at different periods of life ?
2. Give some of the uses of foods.
3. Name foods derived from each of the kingdoms.
4. How are foods classified?
5. Where is salt found in the body ?   In what foods ?
6. Give some of the uses of salt.
7. What proportion of the body is composed of water ?
8. Why is water so important ?
9. What does rain water contain ?   Spring water ?
10. Which form of life flourishes best in spring water ?
11. From what source is drinking water usually obtained ?
12. Is a small amount of lime in drinking water beneficial ?   Why ?
13. How can the germs in water be destroyed ?

CHAPTER III.

THE NITROGENOUS FOODS.

THE organic foods, derived from the animal and vegetable kingdoms, are divided into two classes, depending upon the presence or absence of nitrogen. They are called, therefore, the nitrogenous and the non-nitrogenous foods.

The Nitrogenous Foods. The nitrogenous foods are also called the albuminoids, from their resemblance to albumen. They include such foods as the meats. eggs, and milk. Albumen, as found in the white of eggs. is in nearly a pure state. It forms the chief part of muscle, and is found in nearly all the fluids of the body, including the blood and lymph. As the albuminoid substances are so abundant, entering largely into the formation of the animal tissues and fluids, so their absence from food is soon felt.

No diet can be nutritious, when long continued. unless it contains some of these foods. The nitrogenous foods need not necessarily be from the animal kingdom, as the cereal grains contain nitrogen, but in these the relative quantity is much smaller. The nitrogenous foods are capable of sustaining life much longer than the foods consisting of starch, fat, or sugar. Yet even these foods, when taken alone, are not capable of supporting life. Animals that have been fed on such exclusive diet become enfeebled, refuse food, and finally die. Their continued use, by man, results in a great dislike for them, showing that other materials are necessary to supply all the demands of the system.

It follows, therefore, that while the albuminous foods are the first in importance, yet we must not depend upon one kind of food to sustain life. Foods containing starch and oil are as necessary as are the inorganic substances. The body must be supplied with the same variety of ingredients as that found in its tissues. To deprive it of any one of these will slowly but surely bring on disease and death.

Milk. The "model food" is the name often given to milk; no ideal food can surpass it. As it contains all the food elements necessary for the support of the body, it will, therefore, support life longer than any other single article of diet. It contains a large amount of water, nearly nine parts in ten; a considerable amount of fat, as butter; a sugar, known as milk sugar; minerals; and albumen. The mineral matter consists largely of lime, so essential to the formation of the bones when they are growing rapidly.

The nitrogenous matter consists almost entirely of albumen and caseine. If any acid be added to milk the caseine is thrown down in a coagulated form, and the milk is said to be curdled. The milk curdles spontaneously if exposed to the air for a few hours and in a warm room. This is because of the development of lactic acid in the milk. The action is the same as if the acid were added intentionally, and the caseine thus coagulated. From this coagulated mass, or curd, cheese is made. The fat of milk consists of vast numbers of minute oil drops. These appear, under the microscope, as small round globules floating in water.

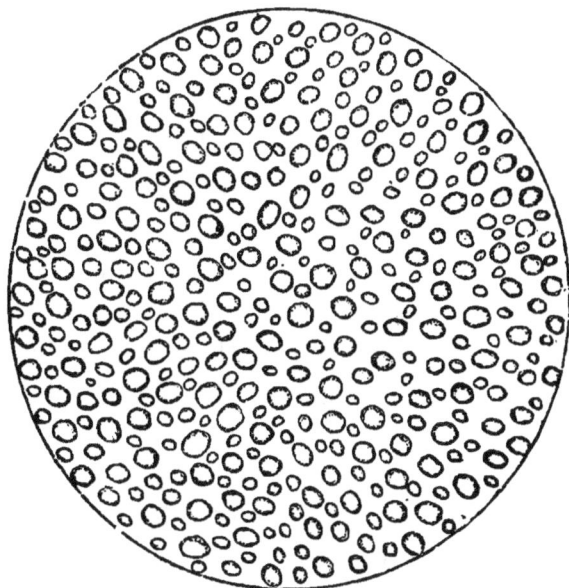

FIG. 3. Milk, highly magnified.

Milk readily absorbs gases ; therefore special care is required to keep it sweet and pure. It must be kept in a room that is very clean and free from all odors. Milk may be the means of communicating disease, either by the use of impure water with which the milk was adulterated, or by the milk having absorbed injurious gases.

Milk should be the principal diet for children, while

for adults it may be used as a drink with the ordinary meals. Warm milk can take the place of the cup of coffee, while cool milk is a good substitute for ice water. Some people claim they cannot use milk, as it disagrees with the stomach and interferes with the action of the liver. These troubles are easily prevented by using only a moderate amount, and by adding to it a small quantity of lime water.

Eggs. Eggs are easily digested and very nutritious. They are the most digestible when soft-boiled in the shell, or when broken into boiling water. The principal differences between the white and the yolk of the egg are these: the white contains no fat, but a considerable quantity of water; while the yolk contains a considerable amount of fat, and but little water. Nearly thirty per cent of the yolk consists of this fat, or yellow oil.

Meats. The meats used for food are rich in nitrogenous ingredients, together with fat and mineral matter. Their nutritious properties and pleasant taste make them very desirable as articles of diet. The meats differ in their digestibility, and in their nutritive value. Beef is regarded as the best meat for general use. When tender beef is properly cooked, it is easily digested and very nutritious. Mutton ranks next to beef, although its flavor is not agreeable to all.

Veal is not easily digested, neither is it so nourishing as either beef or mutton. Pork is not readily digested; the fibers of the lean meat are too compact, and the fat is likely to be in excess. A large class of people, however, eat it freely. To those who exercise much, and who have strong digestive powers, pork appears to do no harm. Oysters are very nourishing, are easily digested, and are very pleasant to those who have acquired a taste for them. Oysters are most easily digested when eaten raw, furnishing in this particular a striking exception to the general rule for all animal substances.

The Cereal Grains. The cereal grains, as wheat, oats, corn, and rice, are most important foods. They consist of nitrogenous material, starch, sugar, salts, and fat. The starch is seen as the white center of the grain, surrounded by a husk. The husk consists of a woody material and is quite indigestible. The nitrogenous portion of the grain is situated between the husk and the starchy center. Wheat flour would be much more nutritious if only the husk, or bran, were removed, and the layer containing the nitrogenous matter and the gluten, oil, and salts retained with the starch. The flour would not be so white, neither would the bread, but it would be more wholesome. The gluten, or the adhesive, jelly-like quality of the cereals, is very abundant in wheat, forming about twenty per cent of the whole grain.

The grains vary in their proportions of nitrogen and starch; but their value as a food does not depend alone upon the amount they contain of any single nutritive ingredient. There should be such a variety of substances as will form the best combination for the nourishment of the body.

Wheat leads all the cereals in nutritive value. It is easily digested and, with the exception of milk, it comes nearest to the standard of a perfect food. It contains but a small proportion of water; has a large amount of starch; is well supplied with nitrogenous material; contains fat; and considerable mineral matter. There is, however, a deficiency in the amount of fat it contains; therefore this must be supplied by putting butter on the bread. The proportion of water is so small that a given bulk of wheat is richer in solids than any other food. Probably the best test there is for a good wheat is the kind of bread it will make.

Bread. Bread is often called the staff of life, because it contains so many nutritious elements, being deficient only in fat. Thus it follows that bread and butter make a very complete diet. In making bread the flour is mixed with water until a dough is formed. Then salt and yeast are added. This is set aside in a warm place until fermentation is well established. The nitrogenous ingredients begin to decompose and act as a ferment on the starch, which becomes, in part at least, changed to sugar. The sugar is further decomposed into carbonic acid and alcohol.

The carbonic acid forms bubbles; these force their way through the dough, or sponge, making the bread rise. The dough is now placed in an oven hot enough to stop fermentation at once. The alcohol is all driven off by the heat, and much of the water also. The bread is then said to be baked. But yeast is not essential to the making of bread. Unfermented bread is made by mixing with the dough a powder composed of an acid and an alkali, so that after the powder is moistened in the bread the acid and alkali form a new compound, and carbonic acid is set free.

In aerated bread the carbonic acid is forced into water, and the flour mixed with this water under pressure. When the dough is heated the carbonic acid expands and makes the bread spongy.

Bread made from unbolted flour is very nourishing, but the presence of the bran makes it difficult to digest, so that it should not be used by persons with weak digestive powers. Hot bread is likely to form a paste in the mouth, and thus its digestion becomes difficult.

Oats, Corn, Rice. Oatmeal contains a large amount of nitrogenous material, ranking in this regard nearly or quite equal with wheat. But it contains a considerable amount of woody or fibrous material, which interferes with the digestion of it and lowers its nutritive value. It also lacks adhesive qualities, so that it cannot be made into bread. Yet it is a wholesome food, and to most persons agreeable.

Corn contains less nitrogenous material than oats, but it has more starch. Rice consists of ninety per cent of starch with scarcely any nitrogenous material. When taken with some albuminous food, as meat of any kind, it is a valuable article of diet; it is easily digested and is also very cheap. Peas and beans are very nourishing, for they contain a good amount of albumen and starch. They would be valuable as food were it not that they consist of such solid matter that they are not easily digested. When used, they should be cooked a long time, and be thoroughly masticated.

Potatoes consist of from seventy to eighty per cent of water; the remainder being nearly all starch, together with a small amount of mineral matter, albumen, and salts. Although they consist so largely of water, yet they are the most generally used of all vegetables. They are used extensively because they can be obtained at all seasons of the year, are very cheap, and agree with most persons. They should never form the exclusive diet;

but, when used with some fat, as butter or meat-gravy, and salt, together with some food rich in nitrogenous matter, they form a most valuable adjunct to the table.

Turnips, cabbage, parsnips, onions, and other vegetables, are added to the list of foods in order to give suitable variety. Their nutritive value is low, and they are not easily digested.

Apples, peaches, pears, and other fruits are valuable foods in many ways. They contain a considerable amount of sugar and mineral matter, while their acids give them a characteristic taste. These acids stimulate the appetite and promote the flow of gastric juice, while the great amount of water they contain serves to quench the thirst. Ripe fruits in their season are most beneficial; while overripe and unripe fruits are often the cause of serious trouble. Much of the danger, however, of unripe fruits is removed by cooking.

QUESTIONS.

1. How are the foods from the animal and vegetable kingdoms divided?

2. The nitrogenous foods include what?

3. Where is albumen found?

4. What is said about supporting life with one kind of food?

5. Of what is milk composed?

6. Is beef nutritious and easily digested? Is mutton? Veal? Pork? Are oysters?

7. Of what do the cereal grains consist?

8. Which is the most nutritive of the cereals?

9. In what is it deficient? How is this supplied?

10. Why is bread a valuable food?

11. What cereal ranks next to wheat?

12. What is said about potatoes as a food?

CHAPTER IV.

THE NON-NITROGENOUS FOODS.

COOKING.

Fats. The non-nitrogenous foods consist of starch, sugar, and the fats. There is a natural craving in the system for these foods, and they cannot be dispensed with for any great length of time without serious injury to the system.

Experiments have proved that some fat must enter into the diet if the bodily health be maintained. This seems to be especially true when the body is growing rapidly. Some individuals do not digest the fat of meats readily, yet they can use butter and milk. Others are able to digest such fatty foods as bacon or the fat of other meats. Fat has great heat-producing power, therefore it is most used where the climate is cold and severe.

It is probable that some of the fat in the body is derived directly from the fat of the food; that is, it becomes absorbed and redeposited in the tissues; but it is equally true that fat may be formed in the body from foods which are without fat. This is proved to be a fact, because the amount of fat, or butter, found in the milk of the cow far exceeds the amount of fat taken as food.

Some persons become very fleshy, while others, with the same diet, remain lean. In many families there is an inherited tendency to accumulate fat in the body. It is probable also that, in some cases at least, more food is taken than is necessary for the normal uses of the body. A deposit of too much fat is attended with inconvenience and no little danger.

Starch. Pure starch is a fine, white powder, consisting of minute granules. When examined under the microscope, the granules are seen to vary in size and form, ac-

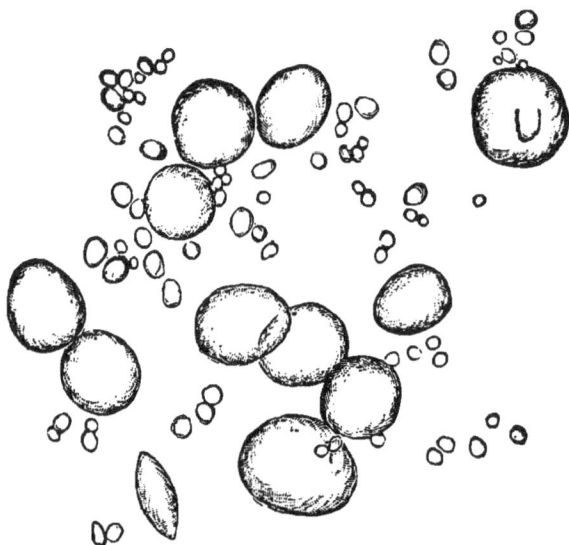

Fig. 4.   Wheat starch, or wheat flour, highly magnified.

cording to the kind of starch. Thus it becomes possible to tell from what vegetable the starch was obtained by the appearance of the granules. The four illustrations here given show these marked differences. Each starch is magnified the same number of times, or about five hundred diameters. The granules are very minute, those of

rice starch being not over $\frac{1}{6000}$ of an inch in diameter.
A study of the starches is very fascinating work to
those who have a microscope. When mixed with water
the granules swell and form a paste; when boiled with
a large amount of water they expand greatly, and can

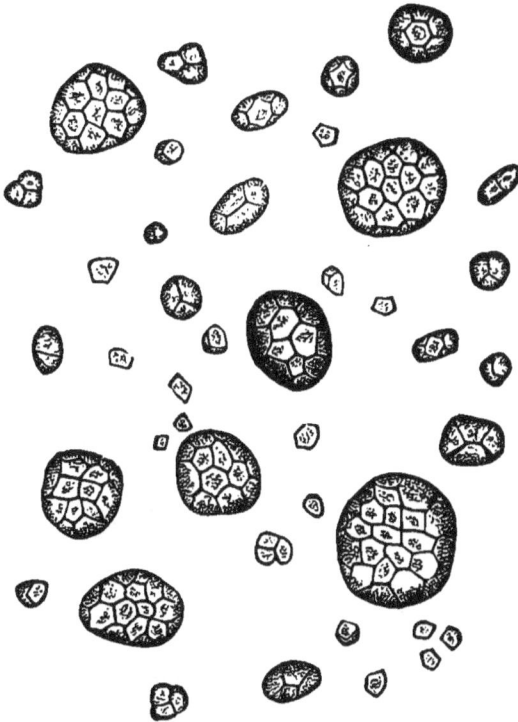

FIG. 5.   Oat starch, highly magnified.

no longer be seen. The destruction of these granules
is a great aid to their digestion. Prolonged cooking
changes the starch into a substance called dextrine.
This is easily changed by the digestive juices into glucose or grape sugar. The brown crust of the bread is
the starch of the flour, changed into dextrine by the
prolonged exposure to heat. It is more easily digested

than the softer parts of the bread. In order that the starch granules may be completely broken up, all starchy foods should be thoroughly cooked. A too exclusive

FIG. 6.   Corn starch, highly magnified.

FIG. 7.   Potato starch, highly magnified.

diet of starchy foods is likely to impair the digestive powers; for the digestive juices are unable to promptly change large quantities of starch into sugar, and the sugar is slowly absorbed if present in too large quantities. This gives rise to the formation of gases and acids, and then dyspepsia follows.

Sugar. Sugar is closely allied to starch, both in its chemical and physiological relations. In the living plant the sugar and the starch represent the same nutritive material, though under different conditions; the sugar is in the form of a liquid, and the starch is in the form of a solid.

There are three principal varieties of sugar, — cane sugar, grape sugar, and milk sugar. Cane sugar is that obtained from the juice of the sugar cane. It is the variety in ordinary use. It is also made from the juice of the maple tree, and is called maple sugar. It is the most soluble and the sweetest of the sugars. Grape sugar is found in great abundance in the juice of ripe grapes. It is generally distributed in the sweet juices of many fruits and flowers. This is the reason it is found in honey, although cane sugar is also present. Grape sugar, also called glucose, is found in some of the animal tissues and fluids, as in the liver and the blood. This is the form of sugar which is made in the body by the digestion of starch. The third variety, the sugar of milk; is found only in milk. Its sweet taste is not very marked.

While it is true that a considerable quantity of sugar is likely to disturb the stomach, yet it is equally true that a certain amount is very desirable. The natural desire for sweet things is so universal in the young that it can indicate nothing else than a demand of the system for this food. The impure and highly colored

candies in the store do not represent a definite amount
of sweet food. It is better to satisfy the craving by
eating ripe fruit. But if candies are desired. let them
be made at home. from pure sugar. There is no danger
of the teeth being injured by the use of sweet foods.
if they be cleansed each day. as they should be under
all circumstances.

The Amount and Kind of Food. No rule can be laid
down saying positively just how much of each kind of
food must be taken : but if the body be in a healthy
condition. an amount should be taken sufficient to satisfy the appetite. The appetite. however. is not always
a safe guide. because by irregular habits. by overeating.
and by eating improper articles of food, it becomes variable. and is then an unreliable test. Persons are liable
to eat too much at a time when some particular article
of diet is especially pleasing, and in this way weaken
the digestive functions.

The quantity of food must vary with the amount and
kind of exercise. A good rule is this : Learn what are
wholesome foods. how they should be cooked. and that
a mixed diet is best : then decide what and how much
to eat. Some persons require only a small amount of
food to keep them in good health. while others require
much more. Yet sickness is caused by overeating. as
well as by eating things which are harmful. Habit and
custom have a great deal to do with determining the
kind and amount of food.

Cooking is absolutely necessary for some kinds of
food, in order that they may be properly digested. It
either softens the food or aids in dividing it into small
particles. It also brings out distinct and agreeable fla-

vors, and thus pleases the taste. If the temperature employed in the cooking be too high, the natural flavors will be destroyed, and disagreeable odors will be produced; while if the temperature be too low, the flavors are not developed.

Cooking coagulates the albumen in the foods; it renders the fatty tissues more fluid; it changes the starchy foods into a pulpy mass, and it breaks up the harder tissues of the vegetables. Thus the foods are softened, and are more easily masticated.

Methods of Cooking. Nearly all the articles of food are cooked before eating, the principal exceptions being oysters and certain ripe fruits.

In broiling, roasting, or boiling meat it is desirable to retain in the meat as much of the nutritive properties as possible. This can be done by applying great heat at first, producing a rapid coagulation of the albumen on the surface. This forms a crust, through which the nutritive juices of the meat cannot escape. Afterwards the cooking should proceed with a less degree of heat, until the meat is cooked to please the taste.

When meats are cooked too thoroughly their natural juices are driven off by the prolonged heat, and their albuminous matter is hard and dry. Such meats are masticated with difficulty and digested slowly, while much that is nutritious is lost. The only argument advanced in favor of such thorough cooking of meat is that it destroys all parasites; but there is little if any danger of these in any of the meats, if pork be discarded.

Broiling is the best method of cooking meat, while roasting is nearly as good. Vegetables and the coarser meats can be made very tender by prolonged boiling, remembering that they · should be placed at once into boiling water, when after the first application of heat the temperature should be considerably lowered. Even potatoes would be more nutritious if boiled with at least a portion of their skins on, as the skins would aid in retaining some of the nutritious materials that would otherwise be lost.

Frying is about the worst of all the methods of preparing meats and other foods for the table. The fat in the meats, or the fat in the frying-pan, penetrates the lean portions and surrounds each particle with a layer of oil. As oil is not digested in the stomach, it follows that the meat with its oily covering must pass out of the stomach before the outer coat of oil can be completely removed. Certain fatty acids are also developed during the frying process which are likely to prove injurious. If food must be fried, the fat should be boiling hot when the articles are put into it, and it should be kept boiling during the entire cooking. Thus by forming a hard outer coat at once, the fat is not so likely to penetrate deeply.

In making soups it is desirable that all the juices be extracted from the meat, — a result just the opposite of the one desired in broiling or roasting. Therefore the meat should be cut into small pieces and placed in cold water at first, and the water gradually allowed to come to a high temperature. In this way no layer of coagulated albumen is formed on the outside, and all the juices are gradually brought out by the water.

Enough has been said to show that healthy, wholesome cooking depends upon well-known laws of chemistry and physiology. A knowledge of these sciences is necessary to fully understand and master the mysteries of this art. But it is one of the happy circumstances of early life that if there is a healthy body to begin with, it does not become necessary to study the articles of food as they are placed on the table. We eat heartily of the good things provided for us; enter into all the labors and pleasures of the day with much earnestness; sleep soundly at night; awake in the morning with rested body and clear brain; and begin again a new day. Thus should it be all through life.

QUESTIONS.

1. What are the non-nitrogenous foods?
2. Are these foods necessary?
3. What especial power has fat?
4. The fat of the body is derived from what?
5. Give a description of starch.
6. What is said of the different forms of starch grains?
7. How does cooking affect starch?
8. What non-nitrogenous food is closely allied to starch?
9. Name the varieties of sugar.
10. Where is cane sugar obtained?
11. Where is grape sugar found?
12. How does cooking affect the food?
13. What is desirable in broiling meats?
14. How can this be accomplished?
15. Why is frying a poor method of cooking meats?
16. How can all the juices be extracted from meat?

CHAPTER V.

ALCOHOL.

THE microscope of modern times reveals to us many wonderful things. It shows the existence of the most minute forms of living objects ; so minute that the highest powers of the microscope and the greatest skill of the observer are necessary to see them. These forms are so minute that they are often spoken of as "micro-organisms." The germs or spores of these micro-organisms float about in the air in a dry state, and are carried with the dust and deposited upon objects everywhere.

Whenever any dead animal or vegetable substance is left in a moist state in moderately warm air, these minute forms enter it and begin at once to take it to pieces. It has been conclusively proven that no putrefaction or decay ever occurs without the presence of some of these micro-organisms, and that they are the cause of such decay. To illustrate the power which they have, it is only necessary to notice any dead animal or vegetable substance. When left to itself such material gradually changes its appearance, becomes softer, the liquids evaporate, the gases escape, and only a little earthy matter is left. This is all the work of these microscopic objects. They have succeeded in completely changing the character of the animal or vegetable substance.

Let us experiment something like this: Boil a piece of meat in some water and then strain the mixture, so that only a clear liquid or infusion remains. Now if this clear solution be boiled and placed in a bottle while at the boiling point, and the bottle tightly sealed, the solution will remain clear indefinitely. A similar result is shown in the canning of fruits; the boiling fruit is placed in cans, which are at once tightly sealed. Why do not these canned fruits decompose? Because the heat applied to them destroys all the micro-organisms, and because no others can gain entrance through the tightly sealed vessels.

Suppose, however, that we open the can of fruit or the bottle of meat infusion. In a few days the liquid will become cloudy, a scum will gather on the surface, and a disagreeable odor will be given off. What is the cause of these changes? Micro-organisms have entered and decomposition is taking place. It is not necessary to add these minute objects to dead substances in order that they decay. The germs or spores that cause decay, as has been stated, float about in the air, though invisible. In fact they are distributed almost everywhere throughout nature.

There are many varieties of these minute organisms, each of which takes its food from a different kind of dead substance, and in so doing, causes that substance to decay or go to pieces. Some of these micro-organisms are called "bacteria," from a Greek word signifying staff or rod. They are so named from their resemblance, as seen through a microscope, to minute staves or rods. Other organisms are known as ferments. Still other organisms are called molds.

Whenever a dead substance is acted upon by these micro-organisms a change takes place which is called fermentation. During this change the elements which compose the fermenting substance are separated. These elements, thus set free, enter into other combinations, forming new substances. The first substance may be very simple and wholesome; but after the elements, as carbon, hydrogen, nitrogen, and oxygen, have been separated by the process of fermentation, they may unite again in a very different manner, producing a highly poisonous compound.

To illustrate this, let us take sugar, which is composed of carbon, hydrogen, and oxygen, combined in a certain exact proportion, and dissolve it in the right amount of water, and add some ferment, as yeast. This ferment, which cannot work on dry sugar, causes the union of the carbon, hydrogen, and oxygen to break to pieces and new compounds to be formed. Some of the carbon and oxygen unite to form carbonic acid gas which escapes in bubbles; while the remaining carbon and oxygen, and all the hydrogen are left so united that a liquid is formed which is called alcohol. Alcohol, therefore, is a result of one kind of fermentation.

There are many kinds of fermentation. For instance, when meat decays, the process is called putrefactive fermentation; it is caused by a certain species of bacteria which do not thrive upon preserves, or other substances not adapted to their growth. The souring of milk is called lactic fermentation; it is caused by another species of bacteria, which changes the sugar of the milk to lactic acid.

The decay of fruits is also caused by micro-organisms, many of which we are familiar with in the form of molds. The microscope shows that these molds are minute plants, the spores or germs of which are floating everywhere in the air.

The spores of other micro-organisms called ferments float in the air, and fall upon the surface of fruits. With them are also the spores of ferments, which are also plants. But between the sweet juices of the fruits, and the ferments there is a sound and complete membrane which we call the skin. While the juice remains in the fruits, the ferments can do no harm. They are harmless if taken as we eat the fruit; but when the juices are pressed from the fruits, these ferments, lodged on the surface, are washed into the liquid. If this mixture be allowed to remain in a moderately warm atmosphere, the spores of the ferments soon begin to grow and multiply with great rapidity.

We know that ripe fruits contain more or less sugar dissolved in the water or juice of the fruit. As we have already seen, the ferments begin at once to take oxygen from the sugar, causing the sugar to break to pieces, and to form two new substances, — carbonic acid gas which escapes in bubbles, and alcohol which remains.

Alcohol is the product of this process of fermentation. Although sugar and water are such harmless foods, yet by the action of the minute ferments, a narcotic poison is obtained from them. Alcohol is called a narcotic poison because of its power to deaden or paralyze the brain and nerves. Like opium, tobacco, and other narcotics, its chief danger lies in its power to set up a continual and ever-increasing demand for itself. It is more dangerous, therefore, than some other poisons, which, taken in quantities too small to cause death, may pass off without doing appreciable injury.

From what has been said, it is evident that alcohol is not a natural part of fruits or grains. With these facts in mind it is easy to understand why beer, wine, and cider are poisonous liquors, although obtained from wholesome fruits and grains. For we must remember that it is a law pervading nature that fermentation entirely changes the character of the substance it works upon. It is another illustration of this law that the fruit juices which have undergone alcoholic fermentation will, if left to themselves in warm air, be entered by another kind of ferment which changes the alcohol to acetic acid. This change, which is called acetous fermentation, is the one that turns cider to vinegar.

Cider. The juice of apples when first pressed from the fruit, consists simply of water, more or less sugar, and a small amount of acid. Thus, in ordinary cider-making, the proper conditions are present for alcoholic fermentation to take place. These conditions are: water, sugar, warm air, and the ever present ferment germs which float in the air, and those which were washed from the surface of the fruit while it was being ground and pressed. Thus it follows that within a few hours after the juice has been pressed from the apples the process of fermentation is well under way.

When obtained from the mill all fresh cider is ordinarily more or less contaminated with the fermented juices which remained in the mill or vats from the previous grinding. It is therefore difficult to obtain cider, even directly from the mill, which does not contain more or less alcohol. Certainly there is no sharp line between sweet and sour cider. If cider does not contain alcohol it is harmless; if it does, it is injurious.

As nearly all so-called "sweet cider" may contain more or less alcohol, and as alcohol even in small quantities has the power to create an appetite for more, and as the delicate tissues of the human system are easily impressed, the testimony of science is strongly against the use of cider in any quantity. It is certainly true that the continuous drinking of old cider makes the person irritable, blunts the finer sensibilities, and weakens the will.

Wine. When the juice is pressed from the grape, the currant, or the elderberry, the ferments on the surfaces of these fruits are carried into the juice and there begin to change the sugar of the juice as in the case of cider. If a small amount of sugar be added to the juice there is then so much more material on hand to be converted into alcohol. For this reason it is true that homemade wines often contain more alcohol than the wines of commerce. Homemade wines may be "pure" as to their freedom from adulteration by drugs, yet the one great poison, alcohol, is present; and for this reason they are pernicious.

Wine is the earliest known and most extensively used of the alcoholic liquors. Its occasional or moderate use is never safe, although it has been so considered by those who were ignorant of the power a little wine has to create an appetite for more. No one can tell how soon such an appetite may be formed, nor how soon it may become an uncontrollable craving, to gratify which is to invite destruction. Ignorance of these facts, or a mistaken confidence in this beverage, has not prevented the evil results that follow its use.

The notion is quite prevalent in many sections of the country that a greater abundance of "light wines" would prevent the use of stronger liquors, and thus diminish intemperance. But the steady growth of intemperance, with its attendant evils of immorality, insanity, and other nervous diseases, in wine-growing countries, where light wines are cheap, plentiful, and universally used, proves the fallacy of this idea.

A recent alcoholic commission, appointed by the French Minister of Finance, reported that alcoholism threatens the people of France with rapid degeneration ; that it is one of the most serious dangers of the times. Not only men but women and children are affected. Mental diseases hitherto unknown have become common.

Dr. J. G. Holland, after spending some time in Switzerland, gave the following testimony as to the effects of light wines there : —

"We have been told in America, and I fully believed it, that if a people could be supplied with cheap wine, they would not get drunk, — that the natural desire for some sort of a stimulant would be gratified in a way that would not only be harmless to morals, but conducive to health. I am thoroughly undeceived. The people drink their cheap wine here to drunkenness. A boozier set than hang around the multitudinous cafés here it would be hard to find in any American city.

"The steady old white-wine topers come into blossom. If you can image a cauliflower of the color of the ordinary red cabbage you can achieve a very adequate conception of faces that are not uncommon in all this wine-growing region. So this question is settled in my mind. Cheap wine is not the cure for intemperance. The people here are just as intemperate as they are in America, and what is more, there is no public sentiment that checks intemperance in the least. The wine is fed freely to children, and by all classes is regarded as a perfectly legitimate drink.

"I firmly believe that the wines of Switzerland are of no use, except to keep out whiskey, and that the advantages of wine over whiskey are not very obvious. It is the testimony of the best people of Switzerland — those who have the highest good of the people at heart — that the increased growth of the grape has been steadily and correspondingly attended by the increase of drunkenness. They lament the planting of a new vineyard as we at home regret the opening of a new grog-shop. They expect no good of it to anybody. They know and deeply feel that the whole wine-growing enterprise is charged with degradation for their country."

Beer. Although beer is made from grain which consists principally of starch, this starch is readily changed to sugar in the process of growing and sprouting. It is only necessary to dissolve out the sugar from grains which have sprouted in order to have a sweet liquid; and this will produce alcohol if the right kind of ferments are added. Therefore, to this sweet liquid, yeast, which is an alcoholic ferment, is added. The ferments, or yeast, begin at once to work upon the sugar, taking oxygen from it and leaving carbonic acid gas and alcohol. The former bubbles up through the liquid and passes into the air; while the latter stays in the beer. Hops also are added, which give the beer a bitter taste.

The food value of grain is almost entirely lost in this fermenting process, while the resulting alcohol has its ever present power to injure and degrade the drinker. Beer so undermines the physical system that beer-drinkers are peculiarly susceptible to disease, and have weakened powers of recovery. A celebrated physician says : " The diseases of beer-drinkers are always of a dangerous character, and in case of an accident they can never undergo the most trifling operation with the security of the temperate."

Distillation. In such mixtures as cider or wine, the alcohol forms only a certain per cent of the whole liquid. To separate the alcohol from the other ingredients it is only necessary to take advantage of the fact that some liquids evaporate much more easily than others. We already know that alcohol is driven off by a low degree of heat, even before the water in the mixture is warm enough to boil. Therefore, if a certain degree of heat be applied to wine, the alcohol will be expelled, as a vapor, before much of the water is given off. Some water, however, escapes with the alcohol, together with some aromatic substances. This is brandy, which consists of about fifty per cent, or one half, alcohol.

Brandy is also distilled from the fermented juice of many other fruits. Whiskey is distilled from the fermented grains. In Ireland it was formerly called by a very characteristic name, the translation of which gives " madness of the head."

Understanding the principle of fermentation and distillation, it is easily seen how a large number of highly intoxicating drinks can be obtained from a great variety of sources. For instance, by steeping roots of plants or trees, thus extracting their flavor, and adding sugar and yeast, it becomes an easy matter to obtain a product of fermentation. Some of these are called homemade beers. The fact that they contain alcohol is their condemnation.

By distillation the alcohol and aromatic qualities can be separated, and powerful drinks obtained. In all parts of the world large numbers of intoxicating drinks are obtained in this way, from sweet grasses, from roots, and from nearly all kinds of fruits and grains. That evil results follow their use in all lands is obvious to even the casual observer.

QUESTIONS.

1. What does the microscope reveal?
2. How do these micro-organisms affect dead animal and vegetable matter?
3. What is necessary for decay?
4. How can we illustrate the power of these micro-organisms?
5. Describe the experiment of boiling meat in water.
6. Why do not canned fruits decompose?
7. Where are the spores or germs found?
8. Name some of the varieties of these micro-organisms.
9. How does fermentation affect a substance?
10. How is alcohol made from sugar, water, and yeast?
11. Give illustrations of some of the fermentations.
12. How do molds cause fruit to decay?
13. Why is alcohol called a narcotic poison?
14. Why is it dangerous?
15. What is said of the fresh juice from fruit?
16. Why is there likely to be alcohol in so-called sweet cider?
17. Why are homemade wines and pure wines not harmless?
18. How does beer affect the system?
19. What are some of the distilled liquors?
20. How are they obtained from cider or wine?

CHAPTER VI.

ADDITIONAL FACTS ABOUT ALCOHOL.

It Creates an Appetite. There are only a few substances which have the power to create an uncontrollable appetite for themselves. The most marked of these are, opium, tobacco, Indian hemp, and alcohol. The peculiarity of these drugs is that when once used there is a strong, almost irresistible desire to use them again. The amount of alcohol in cider, beer, or ale, is sufficient to create this desire. A few glasses of wine will make such an impression on some people that a strong desire for more of the stimulating effect is established. This is due entirely to the presence of alcohol. One drink creates a desire for another, until, as the system becomes more and more under the influence of the alcohol, the desire becomes greater and greater. A small quantity no longer satisfies, and the amount is gradually but steadily increased. Often the entire body is rendered stupid or insensible by the effort to satisfy the demands of this powerful appetite.

In some instances a very little is sufficient to arouse a desire for drink of which the person was wholly unconscious, while another person may take many glasses before he is aware that he has acquired a fondness for it. One of the weighty charges science has to bring against alcohol is, that it has the power to create an ever-increasing appetite for its use. We know of only one way to escape this power: avoid all drinks which contain the smallest quantity of alcohol.

Brandy sauce or wine jelly may seem like very innocent dishes, yet they may be the first step toward creating this appetite. This statement does not come from the imagination of the writer; it is based upon the scientific fact that alcohol has the power to create an appetite for itself, and that this appetite increases as the years go by. It is also based on the testimony of hundreds of people.

The Appetite may be Inherited. It is generally accepted that the continued use of alcoholic beverages, for a considerable period of time, produces marked changes in the whole system. These will be brought out as we study the various organs and tissues. It is safe to assert, therefore, that alcohol produces a disease, known as acute or chronic alcoholism, and also that this disease may cause changes in certain organs and tissues.

Now, we know there are some diseases which are hereditary. We can even go farther than this and state that certain peculiarities and irregularities, the so-called minor ailments, are also hereditary. For instance, in families who have a " nervous history " there are an unusual number of headaches, attacks of indigestion, sleeplessness, and neuralgia. Certain families, for a number of generations, exhibit peculiarities of temper, likes and dislikes, fondness for certain kinds of work, etc., which are doubtless inherited.

The various forms of insanity are striking examples of the power of disease to descend from one generation to another. There is nothing strange or unusual, therefore, in the fact that the love of strong drink may descend from one generation to another.

Not only is the liking for alcoholic liquors a legacy which the drinker's innocent children often inherit, but the irritability of temper, the lack of energy, the weakened will, the untruthfulness, and the propensities to crime which the confirmed drinker brings upon himself, these also may be an inheritance passed on to his children.

It is a Poison. There are some people who claim that alcohol should not be called a poison. What is a poison? Webster says a poison is "any substance which when introduced into the animal organism is capable of producing a morbid, noxious, or deadly effect upon it." Let us study this definition more carefully, and obtain the full meaning of the three qualifying adjectives. Webster defines them as follows: "morbid," not sound and healthful; "noxious," hurtful, harmful, injurious; "deadly," capable of causing death, fatal.

Judged by such analysis, there can be no hesitation in classing alcohol with the poisons. But more than this; in a work before us entitled, "Poisons and Their Antidotes," we notice that one of the poisons given is "alcohol, including brandy, wine, and all spirituous liquors." In addition to this we have the testimony of a large number of noted physicians, chemists, and men of wide experience and learning, all testifying that "alcohol is a powerful poison."

It Shortens Life. The fact that total abstinence lowers the death rate was first shown by a Mr. Neison, of England. He based his conclusions upon statistics which he had been collecting for many years. His figures showed that total abstinence tended to lengthen the probable life of mankind over twelve years.

For many years a large life insurance company kept a separate account of the death rate among the total abstainers. They lately reported that the deaths have been 29 per cent less than among the others insured. Another company asserts that the death rate among total abstainers is from 30 to 40 per cent less than among those not so classed. Then, too, we have the combined statement of twelve presidents of life insurance companies, that "alcohol tends greatly to shorten life."

We must not be confused by noting an occasional exception to this rule. The physiological effects of the continued use of alcohol are such that there can be but one general result, — a weakening of all the powers of man, lowering his ability to withstand disease, producing disease itself, and thereby shortening his days.

It is a Source of Crime. In England, commissions have carefully investigated and reported the effects of the liquor traffic on the morals of men; in our country, eminent judges, who have extensive experience in the trial of criminals and in the investigation of crime, add their testimony; the "London Times" and eminent journals in our own country give the results of special investigations; and the keepers of various prisons send in their reports. From such source of information we learn that four fifths of all the crimes committed in this country are caused, directly or indirectly, by the use of alcoholic drinks.

It is a Cause of Poverty. From such sources as the above, and from official reports of various State, municipal, and charitable organizations, we learn that fully one half of all the taxes paid by the people is required for the support of institutions made necessary by the use of alcoholic drinks. It is estimated that over half a million of persons are so affected, mentally and physically, by the use of alcohol that they are actually unable to labor, or attend to their business. These same authorities tell us that strong drink is the cause of three fourths of the pauperism in this country.

It is Dangerous in Small Quantities. One of the arguments used in favor of drinking beer is this : the amount of alcohol in the beer is so small that no harm follows its use. On this point Dr. S. H. Burgen says : " I think beer kills quicker than any other kind of liquor. My attention was first called to its effects when I began examining for a life insurance company. I passed as unusually good risks five men, who seemed in the best of health, and to have superb constitutions. In a few years I was surprised to see them drop off, with what ought to have been mild and easily curable attacks of disease. Beer had greatly reduced their power to resist disease."

Dr. S. S. Thorn testifies as follows : " Adulterants are not, in my estimation, the important thing; it is the beer itself. Beer accumulates and gathers certain pernicious agencies in the system, until they become destructive. Every man who drinks beer begins to load himself with soft, unhealthy fat."

Dr. Parmelee says : " The majority of beer-drinkers die from dropsy, arising from liver and kidney diseases, a direct result of their habits of life."

The president of a large life insurance company says: " Beer-drinking in every case is peculiarly deceptive at first, and thoroughly destructive at last."

It is not a Food. We have shown that " alcohol and all spirituous liquors " are poisonous. For this reason alone, we should not expect to find them valuable for food. Indeed, there is nothing about alcohol that gives us any idea it has food value. It cannot build up any of the tissues of the body, while it is often the cause of their breaking down. It contains no albuminous, starchy, or mineral ingredients; while the water existing in the alcoholic beverages can be obtained much purer elsewhere. Then again, close observation of its effects on man does not warrant us in believing that it has any value whatever as a food.

This latter statement might not seem true, at first sight, in the case of beer; for heavy beer-drinkers are very likely to be fleshy. But what is the true condition of such a body ? Certainly not one of health. For, while the fat is being deposited beneath the skin so that the whole body looks plump and well kept, the fat is also being deposited in the deeper parts of the body.

In the muscular tissue of the heart and in the cells of the liver this unhealthy accumulation of fat is also deposited. The heart is thereby greatly weakened, so that it cannot do its work well. This causes poor circulation, shortness of breath, and often sudden death. The liver cannot perform its full work, and the usual results of such trouble follow. The blood vessels become weakened from a deposit of fat in their walls, making them much more liable to rupture in the brain, causing death from apoplexy. Then, too, the whole muscular system becomes greatly weakened; and at the same time all the more important organs and tissues in the body lose their proper structure and become more or less changed into fat. Any agent which is capable of bringing about such changes in the system has no place in the list of foods.

Alcohol in whatever quantity or form taken, never aids in the building up of muscle, while its tendency is to destroy; it never furnishes nourishment to the brain, but tends to weaken and dethrone the reason; it never relieves the heart of any work, but often so weakens it that the work is accomplished with great difficulty.

QUESTIONS.

1. What substances have the power to create an appetite for themselves?

2. What is a peculiarity of these drugs?

3. Is a large amount of alcohol necessary to arouse this desire?

4. What is one of the strongest charges against alcohol?

5. How can we escape this power?

6. The continued use of alcoholic beverages produces what?

7. What disease is produced by alcohol?

8. What is said about hereditary diseases?

9. May the desire for alcoholic liquors be hereditary?

10. What proof have we that alcohol is a poison?

11. How does alcohol affect the length of life?

12. Does it cause crime? Poverty?

13. Are large quantities necessary for it to do harm?

14. What is said of alcohol as a food?

15. Repeat the eight sub-heads in this chapter, — the eight charges against alcohol.

CHAPTER VII.

DIGESTION.

Digestion. The substances necessary for the growth of vegetables, as a rule, are taken directly from the soil without change. The vegetables take their food as they find it. The materials which are suitable for the growth of the plant are found at its roots in such a condition that no marked alteration is necessary before they can be absorbed. At least, this appears to be the general rule. But with man this rule does not apply.

Little of our food comes from the inorganic world. It was once organized, and formed a part of the animal and vegetable kingdoms. As a rule, man does not take his food as he finds it. It has to pass through a series of changes before it becomes absorbed into his system. The meats, fruits, and vegetables are taken into the stomach in a solid and insoluble condition. The object of digestion, therefore, is to dissolve and change the food in order that it may be absorbed.

Digestive Fluids must Vary. As the food consists of a variety of substances, with different physical and chemical properties, so there are several digestive fluids, each having its own particular effect. These juices are

. derived from minute glands, situated in the lining membrane of the alimentary canal, and from a few larger glands lying near this canal, with ducts leading directly into it. As the food passes down the alimentary canal, it comes in contact with these juices, and portions of it become liquified, in which condition it can be taken up by the absorbent vessels.

Digestive Apparatus in Different Animals. It is very interesting to study the variety of arrangements of the digestive apparatus in the lower animals. As the different species of animals vary in their habits and in the food they use, so we find a corresponding variation in the anatomy of their digestive apparatus. For instance, in those animals which live upon vegetable substances, the digestive apparatus is very complex. This is necessary because such a large amount of food is required in order that the proper amount of nutriment may be taken from it. The digestible material bears only a small proportion to the entire quantity of food taken. Therefore the alimentary canal must be very large and long.

Take the case of the common fowl, whose food is much more concentrated than that of many other animals. At first, the hard grains are swallowed and held for some time in a pouch called the crop. Here the food mixes with a watery secretion by which it is softened ; as the softened food passes down out of the crop it comes in contact with an acid secretion which is poured from glands in the walls of the tube. The food then passes into the gizzard, which has a very thick, muscular wall. This grinds and crushes the food, aided by the sand or gravel the animal has swallowed, until the mass is reduced to a pulpy consistence. Farther down the canal the food is mixed with more juices, which render it still more soluble. All this complicated apparatus is for the accomplishment of one object, the changing of the food so that it may be absorbed.

In the ox, sheep, and some other animals, there are four distinct stomachs, each performing a different part of the digestive process. The digestive apparatus of man is not thus complex, because his food is comparatively soft, easily made softer by cooking, and also because but little bulk is required to furnish the proper amount of nutritive material. Yet a careful study of the process as it exists in man, will show that it is filled with difficult problems.

The Alimentary Canal. Beginning at the mouth, the alimentary canal extends through the body. It is about thirty feet in length in the adult, and is lined, its entire length, by a soft, velvety tissue called the mucous membrane. In this membrane are minute glands, some of which secrete mucus, while others secrete some of the digestive juices. It is in this canal that the process of digestion occurs. From above downwards are seen the following parts : the mouth, pharynx, œsophagus, stomach, small intestine, and large intestine. A study of Fig. 8 will aid in understanding the location and form of these several parts.

The pharynx extends from behind the mouth about four and one half inches down the neck, where it becomes continuous with the œsophagus. The œsophagus, 1, is about nine inches in length, and extends from the pharynx to the stomach. The stomach, 2, is the most dilated portion of the canal. It lies transversely in the abdominal cavity, and is connected below with the small

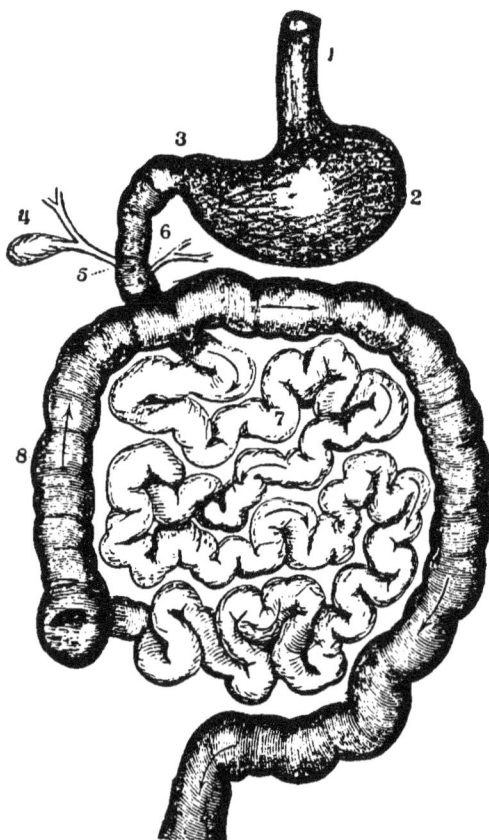

intestine, 7, which is about twenty feet in length. This terminates in the large intestine, 8. It will be noticed that the small intestine occupies the center of the abdominal cavity, while the large intestine passes around the borders of the cavity. All of Fig. 8, except 1, represents that part of the alimentary canal situated below the diaphragm, in the abdominal cavity.

Mastication. Mastication, or chewing, consists in cutting and grinding the food by the teeth. It is purely a mechanical process, yet

FIG. 8. The alimentary canal: (1) The œsophagus; (2) the stomach; (3) the pylorus; (4) the gall bladder; (5) the duct carrying bile to the intestine; (6) the duct from the pancreas; (7) the small intestine; (8) the large intestine.

it is necessary in order that the food may be better prepared for the action of the digestive juices; for the finer the particles of food are, so much the better can these juices act upon them.

One very important result accompanying mastication is the thorough mixing of the food with the saliva. As a result of this, the food is moistened and prepared for swallowing, while at the same time some of its starchy elements are changed into sugar. The solid and semi-solid foods should be chewed very fine. One of the most common causes of stomach trouble is incomplete mastication, a result of too rapid eating. Let nothing pass down the throat that is not crushed and finely divided.

The Teeth. For reasons just mentioned, the teeth are most important aids to perfect digestion. As the habits and foods of animals differ, so do their teeth vary in form and function, in order to best serve particular needs. Fish and serpents, that swallow their food entire, have no need for any cutting or grinding. The function of the teeth

Fig. 9. The skull of a snake.

in these animals is restricted to seizing and holding the food. Therefore their teeth are sharp and curved, with

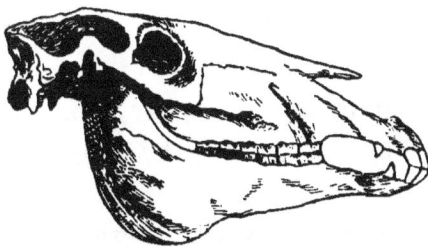
Fig. 10. The skull of a horse.

the points set backward, so that when once the prey is caught it is very difficult for it to escape. In the horse. and allied animals, there are two kinds of teeth, — those in front, the incisors, for cutting off the herbage; and those farther back, the molars, for grinding.

In the gnawing animals, as the rats, mice, and squirrels, the incisor teeth are remarkably developed. Their edges are sharp and chisel-shaped, and they are directly opposed to each other in the upper and lower jaws. They are peculiar also because they grow so that, as the ends are worn away, the tooth is pushed up from its roots, thus keeping its normal length. Sometimes one of these animals has one of its incisor teeth broken off, or injured so that it fails to grow. The corresponding tooth in the other jaw then has no tooth against which to cut, and hence it is not worn away. It keeps on growing, sometimes to an extent sufficient to cause the death of the animal, by preventing it from getting its food.

Two Sets of Teeth. In man the teeth combine the general characteristics already mentioned; but none of them are capable of self-repair, neither do they grow nor alter in shape after they are once formed. As the jaws of a child are not so large as those of an adult, they are not large enough to hold all the teeth of a full-grown person. To compensate for this there are two sets of teeth. One set develops early in life. These teeth are called the milk, or temporary teeth, and are shed during childhood. Another set soon follows, called the permanent teeth.

The first teeth of the temporary set appear about the sixth or seventh month; they appear one by one, until the whole set of ten for each jaw is complete by the end of the second year. In five or six years these loosen and are removed, and the permanent teeth begin to appear. At twelve or thirteen years of age the full set of permanent teeth is present, except the wisdom teeth. These usually do not appear until the person is twenty or twenty-five years of age. The permanent teeth are thirty-two in number, sixteen for each jaw. They are

developed in the jawbones, beneath the temporary teeth.
When they first appear they are of their full size; therefore the permanent teeth of a child twelve years of age
are as large as they ever will be.

Varieties in Shape and Uses. Because the teeth have
different kinds of work to perform they are of various
shapes. Beginning at the middle line of either jaw, and
proceeding backwards, the teeth are placed in the fol-

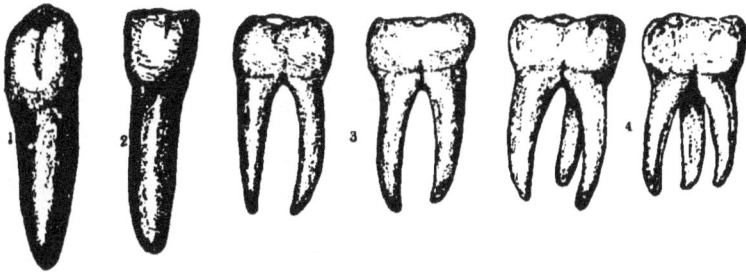

FIG. 11. The teeth of an adult: (1) an incisor, or cutting tooth;
(2) a canine, or eyetooth; (3) molars of the lower jaw; (4) molars of the
upper jaw.

lowing order: two incisors, for cutting the food; one
canine, or eye tooth, pointed and serving the same purpose as the incisors; two bicuspids, or small molars;
and three large molars, for grinding and crushing. The
incisor and canine teeth resemble each other very much,
only if seen in profile it would be found that the crown
of the incisor is thinner and flatter, and has a more
cutting edge. The crown of the canine is more nearly
round, more pointed, and better made for tearing. The
molars of the lower jaw differ from those of the upper,
principally in that they have but two roots, or fangs,
while the upper molars have three.

The Structure of Teeth. Each tooth consists of the
crown, or the part projecting into the mouth; the neck,

or the part surrounded by the gums; and the root, or the part deeply seated in a bony socket. When broken

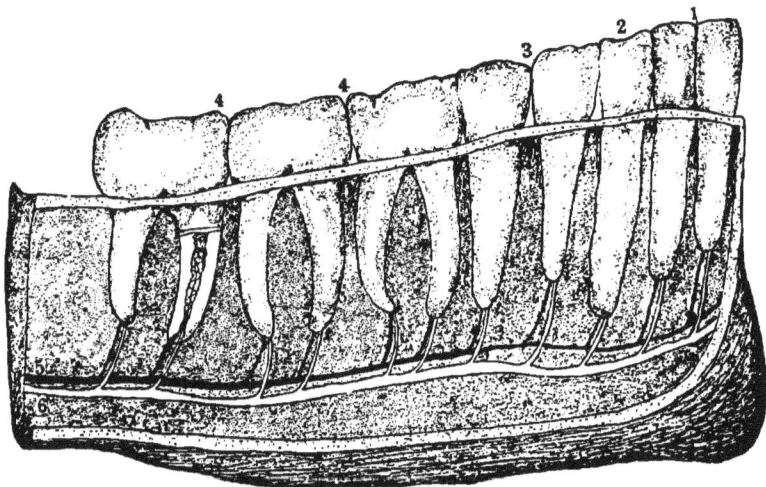

FIG. 12. A side view of the lower jaw with the outer walls of bone removed, showing the teeth in proper place : (1) the two incisors ; (2) the canine ; (3) the two bicuspids ; (4) the three lower molars; (the last molar is sometimes called the wisdom tooth) ; (5) a blood vessel; (6) a nerve.

open, a tooth is seen to be hollow. Fig. 13 illustrates the shape of this central cavity. It conforms to the general outline of the tooth, and thus varies in form for the several teeth. In the living tooth this cavity is filled with nerves and blood vessels, which are held together by a delicate connective tissue. This is called the pulp of the tooth. When inflamed it gives rise to a most intense toothache.

Surrounding the crown of the tooth is the hardest substance in the body, called the enamel; around the root is a thin layer of bone, called cement; but the greater part of the tooth consists of a hard substance, called dentine, or ivory. The dentine surrounds the pulp cavity and extends outwards to the enamel and cement: in structure it is like the tusk of the elephant, harder than bone but not so hard as the enamel.

The figure shows that it is pierced with innumerable fine canals that extend from the pulp to the very outside edge of the dentine. These canals are filled with fibers of living matter which are connected with the cells of the pulp. With the exception of the enamel, therefore, a tooth is a living tissue, having nerves and blood vessels in its center, bone cells in the cement around its roots, and innumerable fibers of tissue penetrating the dentine. With this knowledge it is not strange that decay should make the teeth ache, and that extracting them should cause pain. Yet with all this living matter entering into their structure they cannot repair themselves when injured. They should receive, therefore, daily attention and the best of care.

FIG. 13. Longitudinal section of a tooth : (1) the enamel; (2) the dentine ; (3) the cement; (4) the pulp cavity.

Care of the Teeth. The importance of the teeth to the personal appearance, as well as their relation to the digestive function, is so evident

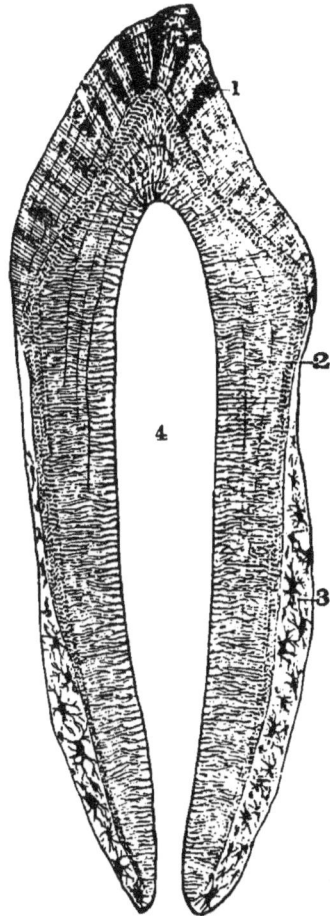

5

that it is a matter of surprise that so little attention is given to their care and preservation. The teeth should be cleaned at least once each day, while it would be much better to cleanse them both morning and evening. Use a small, soft brush and only the best powders, or washes, indorsed by some responsible and well known dentist. Consult a dentist as soon as any cavity is discovered, although a better plan is to have the teeth examined by a dentist every few months.

The Saliva. The saliva is a fluid mainly derived from three pairs of large glands. Mixed with the saliva is some mucus from the mucous glands situated in the lining membrane of the mouth. Two pairs of the salivary glands are situated beneath the tongue and between the two sides of the lower jawbone. The ducts which convey the secretions of these glands open into the mouth just beneath the tongue. The other pair of glands is situated in front of the ears. Each gland lies a little below and directly in front of the external ear. These are called the parotid glands. When they are inflamed they become swollen and painful, and cause the disease known as mumps. The duct from each gland opens into the mouth on the inner surface of the cheek opposite the second molar tooth of the upper jaw.

A drop of the saliva examined under the microscope shows a number of cells that have fallen from the lining membrane of the mouth. As the cells become old they either drop off or are easily removed by the movements of the tongue against them. Fig. 14 illustrates these cells together with others which have escaped from the lymphatic vessels; they are therefore called lymph corpuscles. The vast number of epithelial cells always found in the saliva gives another striking proof that the body is rapidly and continuously changing; for new cells must take the place of the old, as rapidly as they are removed.

Uses of Saliva. The saliva is a constant secretion, although it can be greatly increased by the movements of the jaws, especially when food is being masticated. The saliva is essential in order to keep

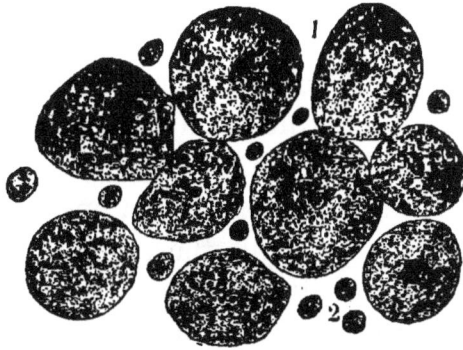

FIG. 14. Cells found in the saliva, magnified: (1) epithelial cells from the mucous membrane lining the mouth; (2) lymph corpuscles that have escaped from the lymphatics of the mouth.

the tissues about the mouth and throat moist. It is extremely difficult to speak if the mouth be dry, as many a young orator knows. The nervousness caused by his first appearance in public temporarily checks the secretion of saliva, and the mouth becomes so dry that speaking is almost impossible.

The principal function of the saliva is to moisten the food, and thus aid in its mastication and solution. It aids greatly, in this way, in swallowing the food, for it is very difficult to swallow anything that is hard and dry, unless first moistened with some fluid. The saliva is also of use, because it dissolves certain substances, and thus enables us to taste them; for solid bodies cannot be tasted. It has also a chemical action on some of the foods; for it is capable of changing starch into sugar. But owing to the short time the food is kept in the mouth, only a small amount of the·starchy foods is thus changed. We shall learn, later, that this important change principally occurs below the stomach.

As one object of the saliva is to furnish moisture, so that the food may be more readily swallowed, it follows that it is not necessary to wash down the food with large quantities of some fluid. As a rule, the saliva furnishes moisture enough, as from one to three pints are secreted each day. While there is no harm in using a moderate amount of drink with our meals, yet large quantities are certainly injurious, especially if they be either very hot or very cold.

The Pharynx. The pharynx extends from the back of the nasal passages downward about four inches, where it becomes continuous with the œsophagus, or gullet. In the upper part of the pharynx, on a line with the floor of the ··   . passages, are the openings of two tubes, called the Eustachian tubes. Each tube extends from the side of the upper part of the pharynx to the middle ear. A disease of these tubes is a frequent cause of deafness.

The pharynx is partly divided from the mouth by a curtain hanging down from above, called the soft palate. It is thus named to distinguish it from the hard palate, which forms the roof of the mouth. From the center of the soft palate there is a prolongation downward, called the uvula, often incorrectly called the palate.

On each side of the throat, below the soft palate, is a tonsil. The tonsils are often enlarged, interfering with speech, and with the swallowing of food. When a severe inflammation of the tonsils occurs, it is known as the quinsy.

Swallowing. Around the lower part of the pharynx are muscles which, by contracting, aid in swallowing. The food is prevented from entering the larynx, or windpipe, which is directly in front of the pharynx, by a valve which shuts tightly down as the food passes over it. Occasionally a small amount of food or drink gets into the air passages, causing violent coughing until it is expelled. The œsophagus is naturally closed, and thus, when a mouthful of food enters it from the pharynx, its muscular walls contract and push the food along until it reaches the stomach.

QUESTIONS.

1. What is the object of digestion?
2. What can you say of the digestive fluids?
3. Describe the digestive apparatus of the common fowl.
4. What animals have four distinct stomachs?
5. Give a general description of the alimentary canal.
6. What is mastication? Why is it necessary?
7. What accompanies mastication?
8. How many sets of teeth in man?
9. How many teeth in the temporary set? In the permanent set?
10. Describe the different shapes and uses of teeth?
11. What is found in the central cavity of a tooth?
12 Where is the hardest substance in the body?
13. Describe the dentine?
14. How should we care for the teeth?
15. Where are the three pairs of salivary glands found?
16 What does the microscope show in saliva?
17. Give some of the uses of the saliva.
18. How does the saliva affect starch?
19. Describe the pharynx.
20. Where is the soft palate? The tonsils?

CHAPTER VIII.

DIGESTION IN THE STOMACH AND INTESTINE.

The Stomach. The stomach occupies the upper part of the abdominal cavity. To its extreme left is the spleen; in front is the abdominal wall; behind are the ribs; below are the pancreas and intestines; above is the diaphragm; and to the extreme right is the liver. When moderately distended it is capable of holding about three pints.

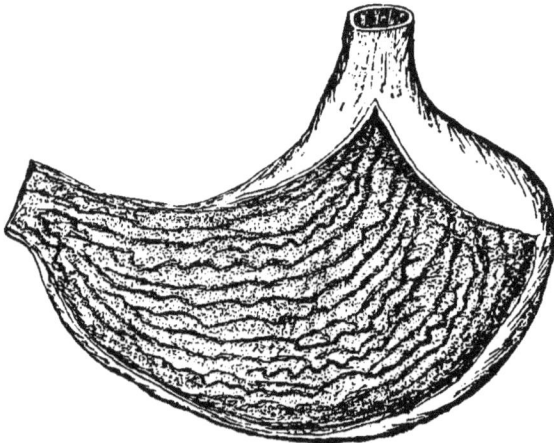

Fig. 15. View of the inside of the stomach. The front walls have been cut away, showing the mucous lining arranged in folds or plaits.

The opening in the right end, through which the food passes out of the stomach into the intestine, is called

the pyloric opening. It is provided with a thick muscular band arranged in a circular manner. This band
is known as the pylorus, so named from a Greek word,
signifying a gate. It is illustrated at 3, Fig. 8. The
object of this band is not to allow the food to pass out
of the stomach until it has been properly acted upon by
the gastric juice.

In the mucous membrane of the stomach are found
vast numbers of minute glands. These are of the nature of tubes, or canals, lined with cells. The cells
secrete a juice called the gastric juice. This is poured
into the stomach, through the openings of the glands,
whenever food is taken.

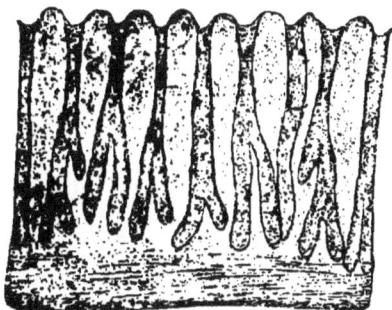

FIG. 16. A cross-section of a small
portion of the walls of the stomach,
slightly magnified, showing the glands.

Fig. 16 illustrates a
section of a small por-
tion of the walls of the
stomach. If this section
were viewed from above,
the minute depressions
would appear as circular
openings. This section
shows that the depres-
sions are the openings of
the glands. Fig. 17 illus-
trates one of the glands, very highly magnified. A careful examination of this figure shows two kinds of cells
in the lower part of the gland. Of such use is the
microscope to the study of physiology that we are now
able to understand the function of these cells, and to
affirm that one kind of cells secretes a certain part
of the gastric juice, while another kind secretes other
parts.

The Gastric Juice. The gastric juice is clear and colorless, in this respect resembling water. But it contains two marked ingredients not found in water; these are pepsin and an acid. When the stomach ·is empty and its mucous membrane thrown into folds, the membrane is of a pale red color, and there is no secretion of the gastric juice. Upon the introduction of food the mucous membrane rapidly fills with blood, becomes bright red in color, and secretes the gastric juice in abundance. Not all foods are digested in the stomach by the gastric juice, for some foods pass out of it unchanged; this is true of the oily and starchy foods.

FIG. 17. One of the glands of the stomach, highly magnified.

The chief characteristic of the gastric juice is its power to dissolve and digest foods of an albuminous nature, as the lean meats, and the white of eggs. If only the proper amount of these be taken, they are completely digested in the stomach; but if more be taken than can be digested by the gastric juice, it passes out into the intestine, where the work is completed by the pancreatic juice. We shall learn that the pancreatic juice possesses the same power to digest albuminous foods as the gastric juice.

Action of Gastric Juice on Different Foods. As has been stated, the albuminous foods, as the lean meats, white of egg, etc., are digested in the stomach. In bread, there is gluten and starch; the former is liquefied and digested in the stomach, while the latter is unaffected. Fatty foods, as the fat of flesh, are affected but slightly by the gastric juice, and only in this way, — the gastric juice liquefies the tissue that surrounds the fat globules, thus allowing the fat to escape in the form of oil drops. But upon the fat itself the gastric juice has no effect; the fat passes out of the stomach unchanged. Milk is coagulated, or curdled, soon after it reaches the stomach. This is due to the coagulation of the casein. The gastric juice digests the casein, but does not affect the oil drops, or fat. The vegetable foods are digested in a similar manner to that already described. The albuminous matters are dissolved and digested, while the oily and starchy ingredients are set free to pass out of the stomach unchanged. As the contents of the stomach begin to pass through the pyloric orifice they consist of digested albuminous foods, starch, fat, and much indigestible material. To this mixture the name chyme has been given.

Movements of the Stomach. As soon as the food reaches the stomach and the secretion of the gastric juice commences, the muscular walls of the stomach begin to contract. These contractions produce such a movement of the contents of the stomach that the food and gastric juice are thoroughly mixed. The food is thus carried back and forth, to every part of the stomach, so that the whole mass may be penetrated thoroughly by the gastric juice, and digestion go on simultaneously in all parts.

Conditions Affecting Digestion. The solid foods are more easily digested if the pieces be minutely divided; therefore swallowing large pieces of food retards digestion. Food should be eaten slowly, that the gastric juice may be formed in sufficient quantity to be thoroughly mixed with it as it is swallowed. A glass of very cold water, hastily swallowed during a meal, might be sufficient to drive the blood from the mucous membrane of the stomach and check the action of the glands. It might require some time for them to recover from the shock of the cold, during which but little, if any, gastric juice would be secreted. This delay would prolong digestion and be quite likely to produce some form of stomach trouble. Too much liquid with a meal dilutes, and therefore weakens the gastric juice. It is much better to quench the thirst before going to the table. Mental and physical fatigue will interfere with digestion; therefore one should never eat a hearty meal when very tired, either from manual labor or from severe mental exercise. A short rest of a few moments before going to the table is a wise precaution in such cases. If the food be eaten slowly there is but little danger of overloading the stomach, but rapid eating is quite likely to result in overeating, which leads to many troubles. Constant eating, or eating frequently between meals, does not give the glands of the stomach time to rest, and an imperfect gastric juice is secreted, indigestion follows, and thus chronic dyspepsia is induced.

There are many varieties of dyspepsia or disordered digestion. Some are easily cured by a proper course of diet, but many are not dependent upon the stomach alone, and are relieved with great difficulty. A few simple rules, well observed, are all that are necessary for those who are yet young and vigorous. A knowledge of the principles of digestion as here given will enable each one to be his own guide in this particular. It is well to keep in mind the fact that the most serious effects produced on digestion are those caused by the use of alcoholic drinks.

The Time Required for Digestion. The time required for foods to be digested in the stomach was first ascertained by a Dr. Beaumont, who experimented upon one of his patients. This patient, Alexis St. Martin, had received a gun-shot wound directly over the stomach. After the wound had healed, there was found an opening which led from the surface of the body directly into the stomach. The opening was usually closed on the inside by a fold of the mucous membrane of the stomach falling over it; but this could be easily pushed aside and the whole process of digestion carefully studied. Since the days of St. Martin, other cases have occurred of such a nature that similar experiments could be made. These have proved that the time required for the stomach to complete its work depends upon many circumstances, some of which have already been mentioned. But in a healthy person it is claimed that digestion is completed in from one to five hours. The following table will give

EASY OF DIGESTION.	h	m	MORE DIFFICULT.	h	m
Rice, boiled	1	00	Potatoes, boiled	3	30
Apples, sweet, raw	1	30	Oysters, fried	3	30
Milk	2	00	Eggs, hard boiled	3	30
Cabbage, raw	2	00	Pork, broiled	3	30
Oysters, raw	2	30	Beef, fried	4	00
Potatoes, baked	2	30	Cheese	4	00
Chicken, boiled	2	45	Cabbage, boiled	4	30
Eggs, soft boiled	3	00	Duck, wild, roasted	4	30
Custard, baked	3	00	Pork, fried	4	30
Beef, broiled	3	00	Pork, roasted	5	15

a fairly correct idea of the average time required for the digestion of several different foods, including those di-

gested in the stomach and those which undergo this
process farther down the alimentary canal. The time
is given in hours, h, and in minutes, m.

The Liver. The liver is the largest gland in the body.
It is situated in the upper part of the abdominal cavity,
just beneath the diaphragm. The greater portion of it is
on the right side of the body, although some of it extends over to the left side. A reference to Fig. 43
will give an idea of its location, as well as its relation
to the diaphragm and to the stomach. The microscope
shows that the liver consists largely of cells. These
cells secrete a fluid, called bile or gall. During the
intervals of digestion the bile collects in a sac, situated
on the under side of the liver, and called the gall bladder. The liver is constantly secreting this bile. The
bile may pass directly from the liver into the intestine
through a duct, shown in Fig. 18, at 7, or it may first
collect in the gall bladder and pass into the intestine
through another duct, at 8. These two ducts, however,
unite into one, at 9, forming one large bile duct. Just
before this duct opens into the intestine it unites with
the duct from the pancreas, and the two terminate in a
common opening, as shown at 10.

The Bile. From two to three pints of bile are secreted
each day. Many experiments have proved that if this
secretion be prevented from entering the intestine, or if
the liver should fail to secrete it, sickness and death will
follow. When the ducts leading from the liver to the
intestine become stopped up in any way, so that the bile
is held back, the blood vessels absorb the bile and carry
it to all parts of the body. This makes the skin yellow,
as in cases of jaundice. The bile aids in the digestion

and absorption of the oily and fatty foods. It moistens the walls of the intestine and renders their contents more liquid. It also does much to prevent the decomposition of food while it is in the intestinal canal.

The Liver Sugar. We know that all the starchy foods must be changed into sugar before they can be absorbed ; therefore all the starch and sugar taken into the body are finally absorbed as so much glucose, or grape sugar. This sugar is carried directly to the liver, where it undergoes a slight change. The liver stores it in its cells, only to give it up again to the blood as the needs of the body demand.

The Pancreas. The pancreas is situated just below and under the stomach, and is about six inches in length. The duct from this gland opens into the intestine, in common with the bile duct, about four inches below the pylorus. In the lower animals the pancreas is known as the sweetbread.

The pancreatic juice has a direct action on all fatty and oily foods. It is the only digestive juice that is able to completely digest the fats and prepare them for absorption. It changes them into a white, opaque emulsion, called chyle. When examined with the microscope, chyle is found to consist of extremely minute particles of fat or oil. The pancreatic juice is also capable of changing the starches into sugar, completing the work which was slightly begun by the saliva. It can also digest albuminous foods, although this is principally done in the stomach.

The Intestinal Juice. Situated in the mucous membrane of the small intestine, are minute glands : these secrete a digestive fluid, called the intestinal juice. It aids in the digestion of the foods, principally the starches and the fats.

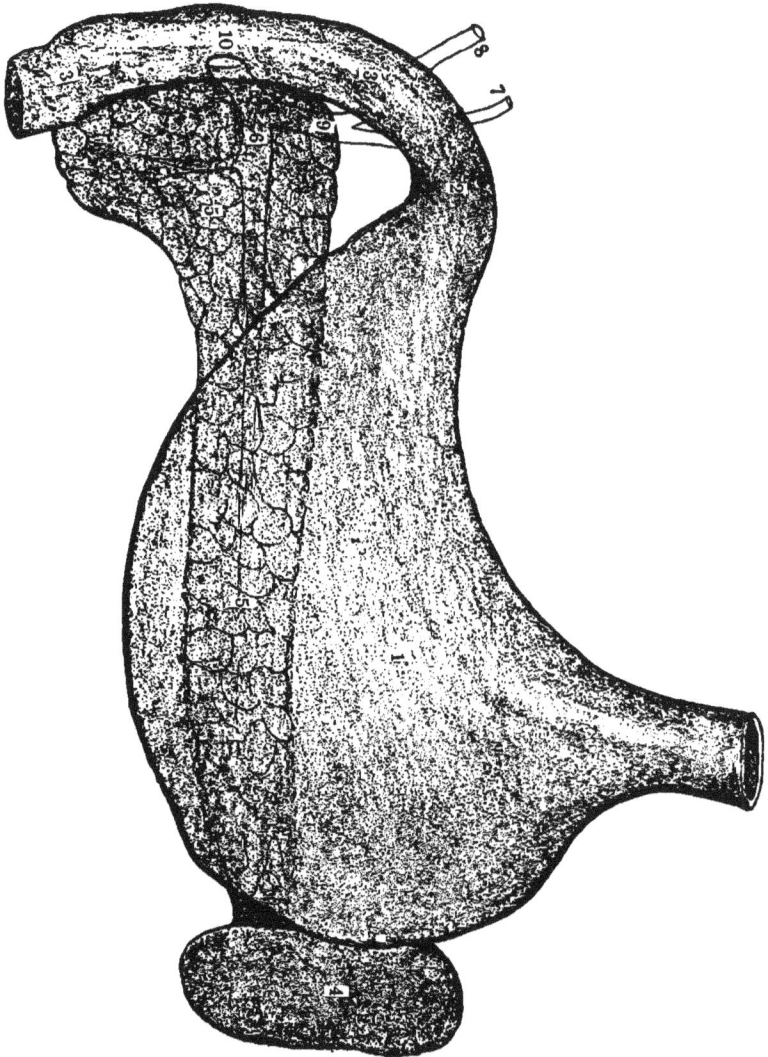

Fɪɢ. 18.   A diagram illustrating the position of the pancreas and its relation to surrounding parts ; (1) the stomach ; (2) the pylorus ; (3) the small intestine ;   (4) the spleen ;   (5) the pancreas ;   (6) the duct of the pancreas ; (7) the bile duct from the liver ; (8) the bile duct from the gall bladder ; (9) the common bile duct, formed by a union of the two bile ducts.   The common bile duct unites with the pancreatic duct, and the one duct thus formed opens into the small intestine at 10.

EFFECTS OF ALCOHOL ON THE STOMACH.

When wine, whiskey, beer, or any other alcoholic liquor comes in contact with the mucous membrane of the stomach, it causes an increased flow of blood to the part. Irritated by the presence of the alcohol the glands throw out an extra quantity of gastric juice. On this account it is considered by some an aid to digestion, but physiologists who have studied the matter say that the presence of the food itself should be a sufficient stimulus, and that the overwork thus thrown upon the glands by the alcohol soon weakens them so that they throw out an imperfect juice. The mucous membrane becomes constantly red or inflamed, and later the glands become smaller and permanent indigestion results.

Such an inflamed condition of the stomach is called gastric catarrh. The inflammation causes an unnatural heat in the stomach, together with a sickening, faint feeling. To quiet the burning sensation and to quench its accompanying thirst, more liquor is taken. This appears to give relief; but the relief is of a most deceptive kind. The alcohol simply deadens for a short time the nerves in the stomach. The apparent temporary relief is to the drinker a sufficient excuse for his continuing its use. Again and again is this experiment repeated, while the inflamed stomach remains a witness to the folly of trying to put out a fire by continually adding more fuel. Persons thus addicted to the use of alcoholic drinks many times make earnest efforts to do without them ; but the craving of the inflamed stomach, the unnatural thirst, and the strong appetite, all appeal for more drink.

Now, what is the condition of a man under such circumstances? He is ill, suffering from an inflammation of the stomach and a disordered liver. Weakened in body and in mind by the disease, he should be treated as a sick man, and by a skilful physician who would not make matters still worse by prescribing alcohol.

The long-continued use of alcoholic drinks in large doses makes most marked changes in the structures of the stomach. The blood vessels become permanently distended with blood, thus interfering with proper circulation. As a result, some parts of the mucous membrane become so diseased that they break down, forming ulcers. This is one of the last and most severe effects. The ulcers give rise to the most severe pain, and this often leads to the use of some form of opium.

In conclusion, we sum up the effects of alcoholic beverages on the stomach as follows: —

A small dose, when not repeated. Increased flow of blood to the stomach; digestion retarded by the action of the alcohol on the gastric juice; probable recovery.

A large dose, not repeated. Increase of the above effects; acute inflammation of the stomach; digestion temporarily checked. Recovery after a few days, or the inflamed condition may remain a long time.

A very large dose, not repeated. Increase of the above effects; occasionally sudden death. Recovery always slow, and often incomplete.

Small doses, when often repeated. A slow, chronic inflammation of the stomach, causing dyspepsia. Recovery impossible while alcohol is used; but possible if its use be stopped.

Large doses, long continued. Increase of the above effects; the glands become reduced in size; the blood vessels are permanently enlarged; sometimes ulcerations occur. Complete recovery doubtful, even if the use of alcohol be stopped.

Very large doses, often repeated. Increase of above effects; stomach retains but little if any food; great pain; opium used; ulcerations deeper; hemorrhages; death.

Effect on the Liver. The secretion of bile and the storing up of the liver sugar can only be carried on properly in a healthy liver. Each cell must be ready to do its part. Alcohol makes marked changes in the liver, resulting in great impairment of digestion, and, therefore, of the whole system.

It is probable that nearly all the alcohol taken as a drink is absorbed while in the stomach. From the stomach, the blood vessels carry it directly to the liver, where it does immense harm. It may cause the liver to become large and fatty, as it does in those who drink beer. In these cases the microscope shows that each cell has become filled with minute globules of fat. Such a condition is represented in Fig. 19, at B. These fat globules cannot make bile, neither can they store up the liver sugar; therefore the liver becomes a great mass of fatty tissue, unable to do the work assigned it.

After the long-continued use of alcoholic beverages, especially whiskey, brandy, and gin, the liver undergoes other changes. It becomes greatly reduced in size and much too hard. Such a liver is so characteristic of alcohol poisoning that medical authorities have given it the distinct name of the " drunkard's liver." The

microscope shows the cells of such a liver to be much reduced in size, and otherwise changed in appearance.

The liver is probably one of the first organs to be seriously affected by alcohol.   If it cannot perform its func-

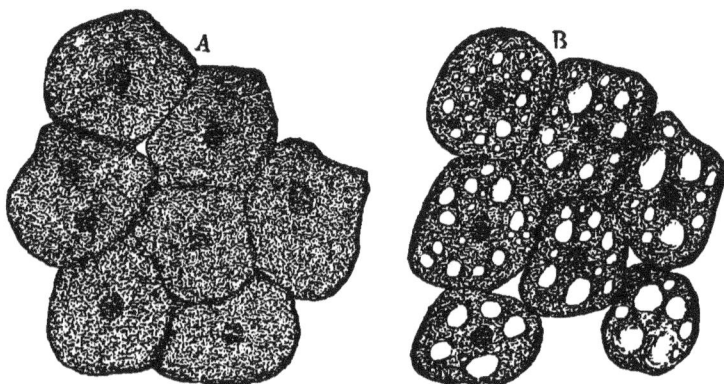

Fig. 19.   (A) Liver cells, highly magnified, from a healthy liver.
(B) Liver cells, highly magnified, from a fatty liver, the oil globules
take the place of healthy liver substance.

tions properly the health of the whole body will, sooner
or later, become seriously affected.   Alcohol does not
simply interfere with the functions of the liver; but it
strikes deeper, and actually changes the very structure
of that important organ.

Effects of Tobacco on Digestion.   Digestion is often impaired in those who use tobacco.   By chewing tobacco the
salivary glands are constantly overworked, so that when
the saliva is most needed, at mealtime, an insufficient
amount is furnished.   This necessitates the use of some
other liquid to moisten the food ; therefore an excessive
amount of water, tea, or coffee is used.   The more general effects are of a secondary nature.   The nicotine is
absorbed in sufficient amount to affect the nervous sys-

tem, giving rise to a kind of indigestion called nervous dyspepsia.

Opium. Opium is such a powerful narcotic that moderate doses of it are sufficient to check nearly all the phenomena of life, while large doses will cause death. Even a moderate dose taken just after a meal may completely arrest digestion. The use of opium often follows the prolonged use of alcoholic beverages, for it quiets the pain and restlessness of the diseased body, and it dulls the intellect so that the sufferings of remorse are not so keen.

Tea and Coffee. Both tea and coffee are likely to interfere with the action of the stomach, especially if taken strong and in large quantities

QUESTIONS.

1. Name the parts surrounding the stomach.
2. Where is the pylorus, and what is its object ?
3. Describe the glands in the mucous membrane of the stomach.
4. Describe the gastric juice.
5. What is the chief characteristic of this juice ?
6. What other juice has the same power ?
7. What is the action of the gastric juice on lean meat? white of egg? bread? fatty foods? milk ?
8. What is the object of the movements of the stomach?
9. Give some of the conditions affecting digestion.
10. Name some foods easy of digestion ; some more difficult.
11. Give the location of the liver.
12. What does the liver secrete ?
13. How does this secretion reach the intestine?
14. What is said about the liver sugar?
15. Give the effect of pancreatic juice on the various foods.

CHAPTER IX.

ABSORPTION.

Definition of Absorption. It has been stated in a previous chapter that the saliva and pancreatic juice change the insoluble starchy foods into the soluble glucose, or grape sugar; that the lean meats, eggs, and other albuminous foods are digested by the gastric juice; and that the fats are changed by the pancreatic juice. If our foods could be absorbed in their natural state, this complicated work of digestion would be unnecessary. But we know that the foods must first be liquefied and changed, before they can be taken up by the proper vessels, and carried to the various parts of the body.

Absorption, therefore, is the process by which the digested food passes from the alimentary canal into the blood vessels and lymph vessels.

Absorption from the Stomach. The water that is taken as drink, and also that found in the food, is largely absorbed by the blood vessels of the stomach. Such mineral salts as are soluble in water are here absorbed; as are also the various alcoholic solutions. There is a slight absorption of the foods, as they are digested in

the stomach. Generally speaking, however, with the exception of the various drinks, both simple and alcoholic,
there is little absorption by the blood vessels of the
stomach. The digested foods are principally absorbed
in the small intestine.

Structure of the Small Intestine. The outer walls of the
small intestine are composed of involuntary muscle
which is directly continuous with that forming the walls
of the stomach. Within this muscular wall, and attached to it, is a mucous membrane which lines the
whole of the small intestine. This mucous membrane
is arranged in folds, or plaits, which pass around and
transversely to the canal. Some of the folds are nearly
two thirds of an inch in depth at their broadest part,
though most of them are smaller. These folds retard
the passage of food along the intestine, and also increase the surface for absorption. Projecting from
these folds, and covering their inner surface are very
minute elevations, called villi. They are from $\frac{1}{30}$ to $\frac{1}{50}$
of an inch in length and they hang down toward the
center of the canal like so many minute fingers. They
give to the mucous membrane its velvety appearance.
It is estimated that there are fifteen or twenty million
of these villi in the small intestine.

Fig. 21 illustrates the general arrangement of the
various parts as seen in a cross section of the small intestine. The outer wall is thick and firm, composed of
strong muscular tissue. Within this are represented
the villi, hanging down toward the center of the canal.
Two kinds of vessels are illustrated in the drawing, —
the blood vessels and the lymph vessels; the latter
are also known as the lymphatics, or the lacteals.

A careful study of one of these villi is necessary for a clear understanding of the subject of absorption. At Fig. 20 is a single villus, highly magnified. Each of those represented in Fig. 21 would appear the same under an equally high magnifying power; in fact, this single villus may be taken as a representative of the

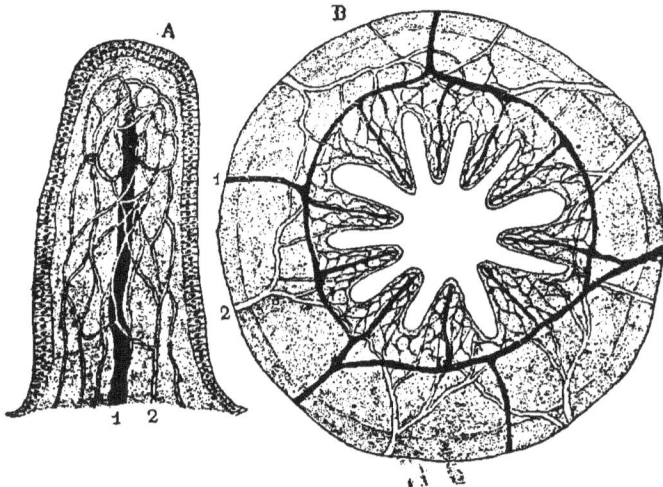

FIG. 20. A villus of the small intestine, magnified: (1) central lymphatic; (2) blood vessels.

FIG. 21. Diagram of a cross section of the small intestine : (1) lacteal, or lymphatic vessels ; (2) blood vessels.

twenty million found in the body. Each villus is surrounded with a layer of minute cells. Directly in the center, 1, is a darkly shaded vessel; this is the lacteal, or lymphatic vessel. It begins at the free end of the villus and unites with the lacteals from the other villi, as clearly seen in Fig. 21. Surrounding the central lymphatic are numerous capillary blood vessels. A reference to Fig. 21 shows that the capillaries from several villi unite to make the larger blood vessels, at 2 The villi, therefore, are found to consist of two kinds

PLATE II.

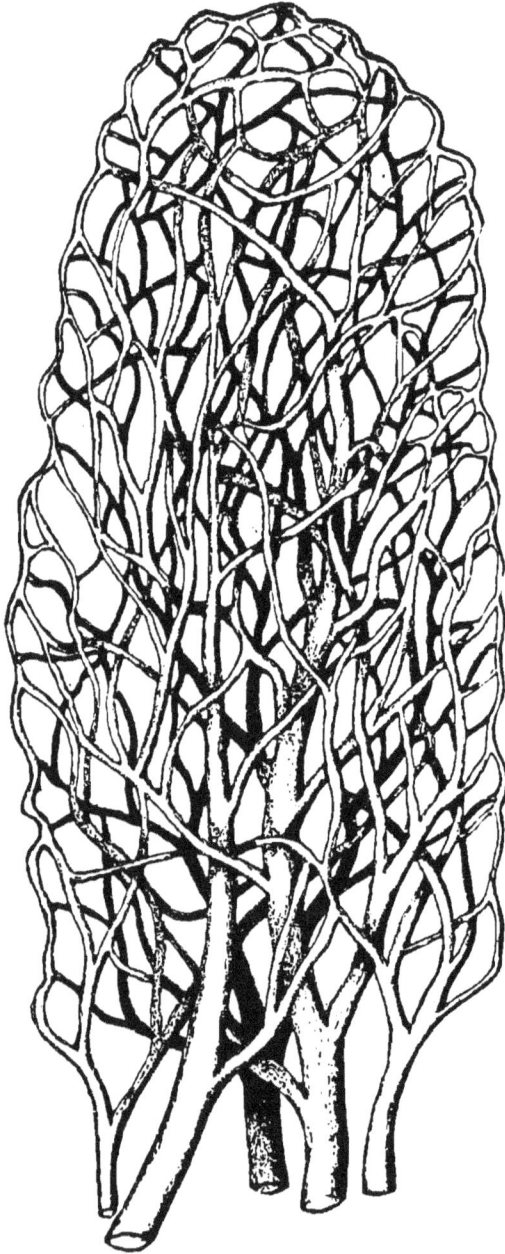

of vessels, surrounded with a layer of cells. Plate II. gives an idea of the vast network of blood vessels found in each villus. No pen can either picture or describe this most beautiful and most complicated arrangement of vessels.

Absorption from the Small Intestine. No better illustration could be given of the value of the microscopical study of the tissues, as an aid to the study of physiology, than the case before us. It is now easy to understand how the digested food can pass from the intestinal canal to the blood vessels and lymphatics. The digested foods can easily pass through the thin layer of cells surrounding the villi, and through the thin walls of the vessels within. The only question to be decided is what kind of food each system of vessels absorbs.

The central lymphatics, or lacteals, of the villi are especially concerned in the absorption of the digested fatty foods. The blood vessels of the villi absorb the other digested foods, as the glucose and the albuminous foods. This is the general rule, although it is a fact that each system of vessels may absorb all kinds of digested food. Water and many watery solutions are freely absorbed while in the small intestine.

The Portal Vein. The blood vessels of the villi unite with other blood vessels from the stomach to make a large vein, called the portal vein. This vein goes directly from the stomach and intestine to the liver. It carries the digested foods, taken up by the blood vessels, at once to the liver. At the proper time this food will pass from the liver, through certain veins, directly to the heart. From this central point it is soon sent out to all parts of the body.

The Lymphatics. The lymphatics of the body consist of the lymphatic glands and the lymphatic vessels. The lymphatics found in the villi are often called the lacteals, because here, when filled with the chyle or digested fats, they look white; hence lacteal, from the Latin word *lac*, milk.

The lymphatic vessels are found in nearly every organ and tissue in the body. They contain a clear, colorless fluid, called the lymph. The general object of the lymphatics is to collect the fluid that is in the tissues of the body and carry it back to the blood. The blood brings nourishment to the tissues ; this nourishment is in the form of a watery fluid which passes through the walls of the blood vessels and penetrates between the fibers and cells of the tissues. From this fluid the tissues take such ingredients as they need for their nourishment and growth. The fluid, therefore, soon has its very life, its most nourishing ingredients, taken from it by the tissues ; nothing is left of it but useless material. At the same time that the tissues are taking up nourishment, they are also throwing off worn-out material ; this forms part of the fluid that is in the tissues. As already stated, the lymphatics gather up this fluid, so that, eventually, its harmful ingredients may be cast out of the body.

All along the course of the lymphatic vessels are minute glands, called the lymphatic glands. The lymphatic glands of the neck sometimes get inflamed and swollen so that they can be felt beneath the skin as minute kernels.

The Lymph. After a meal containing fatty foods, the lymph in the thoracic duct changes from a clear, watery fluid to a milk-white color. This is due to the fact that the lymphatics of the villi take up the digested fats, or chyle, and carry them directly to the thoracic duct. As the digested fats are of a milky color, so the contents of the thoracic duct become of a like color; this only lasts, however, while the chyle is being absorbed. All the other lymphatics are constantly filled with the colorless lymph. As it is the business of the lymphatics to collect a liquid which is in the tissues and carry it eventually to the heart, so it follows that the flow of lymph is always from the periphery toward the heart, being in this particular like the venous blood.

As the lymph flows through the lymphatic glands on its way to the heart, it meets with some changes, chiefly in the addition of the lymph corpuscles. These are minute bodies identical with the white corpuscles of the blood. In fact, they become the white corpuscles as soon as they are brought to the blood by the lymphatic vessels.

The Thoracic Duct. Directly in front of the spinal column lies the thoracic duct. It is from eighteen to twenty inches in length, in the adult, and is about the size of an ordinary slate pencil. This duct carries the greater part of all the lymph and chyle into the blood; while it is the central, large vessel for all the lymphatics of the body. There are numerous valves throughout its entire length, so arranged that they completely prevent the lymph and chyle from falling towards its lower part. The duct begins in the lower part of the abdominal cavity by a triangular enlargement, and then passes up through the diaphragm. When near the heart, it makes a sharp curve and empties into a large vein, beneath the left collar bone. This vein carries the lymph directly to the right side of the heart. Thus the lymph enters the general circulation.

The lymph from the upper part of the right side of the body reaches the circulation through another lymphatic duct of small size. It empties into a corresponding vein beneath the right collar bone.

Colored Plate. A study of Plate III. will aid in understanding the relation of certain parts to each other: (1) the beginning of the thoracic duct; (2) the termination of the duct in the large vein, near the heart; (3) the right lymphatic duct; (4) the spinal column; (5) the large vein which empties into the right auricle, — it corresponds to 2, figures 30 and 31; (6) the aorta; (7) the artery which carries blood to the right side of the head; (8) to the left side; (9) arteries supplying the organs and also the tissues of the abdominal cavity; they are also represented in Fig. 36.

Review. From what has been said it now becomes possible to trace the foods from the time they are taken into the mouth until they enter the blood. Take the three representative foods : lean meat, starch, and fat : —

First, mastication, or chewing; second, insalivation, or mixing with the saliva ; third, deglutition, or swallowing; fourth, stomach digestion, for the albuminous foods; fifth, intestinal digestion, for the fatty and starchy foods ; sixth, absorption ; seventh, albuminous foods and glucose carried to the liver by blood vessels ; eighth, fatty foods, the chyle, carried to the blood by the lacteals.

A study of Fig. 22 will aid the memory in fixing the facts already stated. Begin with the four villi at the

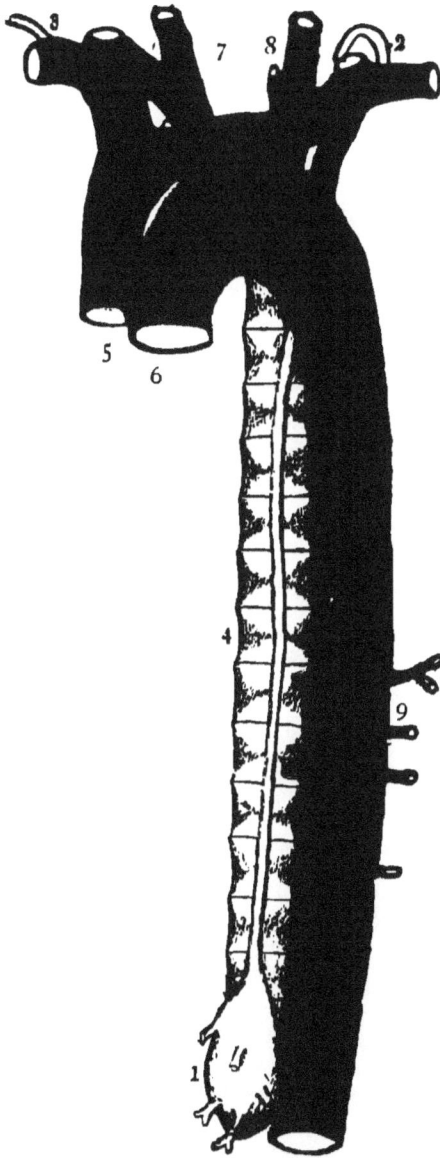

PLATE III.

right of the intestine: suppose the central lacteals are filled with chyle, or digested fats. The four lacteals unite to form a larger vessel, L, which passes through a lymphatic gland, G, and empties into the dilated beginning of the thoracic duct, D. The chyle then passes up the duct in the direction of the arrowheads, until it enters the large vein which leads directly to the right side of the heart. Consult also the colored plate.

Suppose again that the blood vessels of the villi at the left, 2, are filled with absorbed foods derived from the lean meats and starches. The ves-

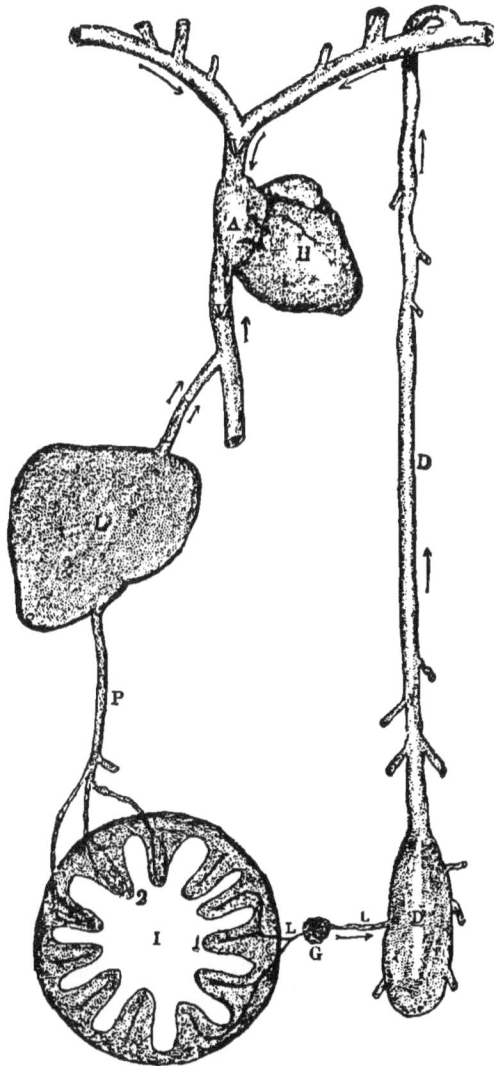

FIG. 22. Diagram illustrating the course of the absorbed foods. I, intestine . (1) villi with central lacteals ; (2) villi with blood vessels. L, lymphatic or lacteal vessels. G, lymphatic gland. D, thoracic duct. P, portal vein. L, liver, at the left of the figure. V, vein. H, heart. A, right auricle of heart.

sels soon unite to form the portal vein, P, which carries the food directly to the liver, L. From the liver it can pass through the veins, in the direction of the arrow-heads, to the right side of the heart.

Fig. 22 is a diagram to illustrate the course pursued by the digested foods, from the intestinal canal to the heart; while the colored plate is a correct representation of the relation of the various parts.

QUESTIONS.

1. Give a definition of absorption.
2. What food is largely absorbed while in the stomach?
3. Is much of the food absorbed while in the stomach?
4. Where does absorption principally take place?
5. What composes the outer walls of the small intestine?
6. What is within this wall?
7. How is this membrane arranged?
8. What is the object of these folds?
9. What projects from these folds?
10. Describe a single villus.
11. What is said of the number of these villi?
12. What foods are principally absorbed by the lacteals? By the blood vessels?
13. Where is the portal vein?
14. Why are the lymphatics of the villi often called lacteals?
15. What is the principal object of the lymphatics?
16. What minute bodies are found in lymph?
17. These corpuscles become what?
18. Describe the thoracic duct.
19. Of what use are its valves?
20. Describe the course of the digested fats from the intestine to the heart. Also the course of the digested lean meats.

CHAPTER X.

THE BLOOD.

General Description. The blood is often called the vital fluid, because it affords life to all the tissues. A sudden loss of much blood causes great weakness, while if the amount lost be considerable, death may result. The more rapid the loss of blood, the more dangerous it is; while if the amount be moderate and the bleeding slow, the loss is better borne and sooner made up.

In a few cases human lives have been saved, after great losses of blood, by having fresh blood from a person or animal injected into the veins. This is called the transfusion of blood. Three or four ounces only are injected, and a few cases are on record where this small amount was sufficient to restore health to persons who were very near death. Many years ago a few remark-able cases of recovery made some people believe that the transfusion of the blood of young persons into the veins of the aged would bring back youth and vigor to the lat-ter. But a number of deaths resulted from this being improperly done, and in some countries the operation was forbidden by law. It is rarely employed at the present time.

Although the blood is very generally distributed through the body, yet there are parts in which it is not found, as the hard parts of the teeth, the hair, the nails, the outer layer of the skin, some parts of the eye, and most of the cartilages. These are nourished by absorbing the fluids which escape from neighboring blood vessels. It is estimated that about one twelfth of the weight of the body is composed of blood.

Medium of Exchange. The blood receives a large amount of new material from the digested food, and a supply of oxygen from the air in the lungs. It carries these fresh supplies to the various organs and tissues; gives them up where they are needed; and receives in exchange, carbonic acid and other products of waste. From this it is seen that the blood always contains both new and old material; new material on its way to build up tissue, and old material on its way to the kidneys, the lungs, and the skin. The blood brings useful material and takes away useless material. It is a great medium of exchange between the outside world and the tissues of the body.

Composition of the Blood. Blood appears to the un-aided sight as a thick, opaque, red fluid. But the microscope shows that it consists of two parts : a transparent, nearly colorless fluid, called the plasma ; and a large number of minute bodies floating in the plasma, called the blood corpuscles. The plasma looks very much like water, yet it holds in solution many important substances.

The Blood Corpuscles. The blood corpuscles are of two kinds, the white and the red. The white corpuscles, as their name indicates, are without color. They are spherical bodies, averaging about $\frac{1}{2500}$ of an inch in diameter. They are a trifle larger than the red corpuscles, but not so abundant, the average number being one white corpuscle to about three hundred red, although this is subject to variation even in health. The white corpuscles are capable of changing their form by a flow of their substance in various directions. This is after the manner of the amœba, as illustrated in Fig. 2 ; hence the changes in their shape are called the "amœboid movements." The white corpuscles are identical with the lymph corpuscles. When, in cases of illness, the person becomes weak and pale, there is generally an increase in the number of the white corpuscles, and a decrease of the red.

The red corpuscles of human blood are circular bodies,

FIG. 23.   FIG. 24.

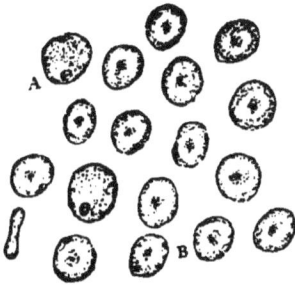 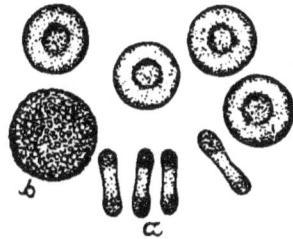

Fig. 23. Human blood, highly magnified : (A) the white corpuscles ; (B) the red corpuscles.

Fig. 24. Human blood, more highly magnified: (a) the red corpuscles ; (b) a white corpuscle.

slightly hollowed towards the center. Fig. 23 shows a number of these bodies, shaded in the center to give the correct impression that they are concave. One corpuscle, at the left of the figure, is seen on the edge.

showing that it is biconcave, or slightly hollowed on either side. The red corpuscles exist in vast numbers. It is estimated that in a minute drop of blood there are over five millions of them; while in a medium-sized person there are not less than twenty-five billions, — too vast a number for the mind to comprehend. Fig. 24 represents a few red corpuscles and one white corpuscle, very highly magnified. Four red corpuscles are seen resting on the side, while four are on the edge.

Function of the Red Corpuscles. The most important ingredient of the red corpuscles is their coloring matter, called hæmoglobin. This substance has a strong liking, an " affinity," for oxygen. So strong is this affinity that when the blood flows through the lungs the coloring matter takes oxygen from the air which it finds there. The red corpuscles thus become loaded with oxygen. The blood soon leaves the lungs, and flows to the most distant tissues, which are in great need of this oxygen. But the tissues exert a stronger affinity for the oxygen than even the hæmoglobin, and thus the latter is obliged to give up the oxygen. For this reason the red corpuscles are called the oxygen-carriers; for their great object is to carry oxygen from the lungs to all the various organs and tissues of the body.

Arterial and Venous Blood. As soon as the red corpuscles receive their fresh supply of oxygen in the lungs they become bright red in color, making the whole blood a bright scarlet. This bright-colored blood is called arterial blood. It is found in the arteries, or in those vessels which carry blood from the heart to the distant tissues. In one place arterial blood is found in the veins, — the pulmonary veins, — which carry the blood from the lungs to the left side of the heart.

When the blood is passing through the smallest vessels, the capillaries, it gives up its oxygen to the tissues. Deprived of its oxygen, the hæmoglobin of the red corpuscles becomes much darker in color, therefore the whole blood looks darker. This darker-colored blood is called venous blood. It is found in all the blood vessels which carry the blood from the tissues back to the heart. It is also found in one artery, — the pulmonary artery, — which carries the blood from the right side of the heart to the lungs. The rule is that the arteries contain the bright arterial blood, and the veins the dark venous blood; but to this there are the two exceptions already given, the pulmonary artery and the pulmonary veins.

Oxygen and Carbonic Acid. The air we breathe consists principally of two gases, — oxygen and nitrogen. The oxygen is essential to all life. Without it we should soon die. All parts of the body use it. The tissues are constantly demanding it and countless numbers of corpuscles are continuously and rapidly at work distributing it throughout the body. We know that a substance called carbon forms a part of all the tissues. When the oxygen reaches the tissues, it unites with their carbon, forming carbonic acid. This is a poisonous gas, and the body must cast it off as soon as possible; therefore it mingles with the plasma of the blood and is soon carried to the lungs, where it escapes from the body.

Arterial and Venous Blood Compared. From what has been said we are able to place in a more concise form the differences between arterial and venous blood : —

7

Arterial blood contains the more oxygen.

Venous blood contains the more carbonic acid.

Arterial blood is of a bright scarlet color.

Venous blood is of a darker, nearly purple color.

Arterial blood parts with its oxygen in the capillaries.

Venous blood parts with its carbonic acid in the lungs.

Arterial blood contains substances for the nutrition of the tissues.

Venous blood contains the worn-out materials from the tissues.

Coagulation. Soon after blood has escaped from a blood vessel, it thickens to a jelly-like mass. This is called the coagulation or clotting of blood. It is one of the wise provisions of nature, so that our lives may not be sacrificed as a result of some slight cut or wound. The lower animals are often severely bruised and wounded, and they may even lose a portion of a limb, laying bare large blood vessels, without fatal results. This is because the blood soon clots, forming a solid mass at the openings of the vessels and preventing any further escape of blood. Whenever any vessel of considerable size is ruptured, it is advisable to aid nature by checking the flow of blood for a short time, in order that the clot may be well formed. This is accomplished by pressing on the part, or by placing a fine thread around the ends of the ruptured vessel.

Blood seldom clots while in the blood vessels of the living body. It never does so unless some disease or some unusual condition be present. As a result of a diseased condition of the walls of the blood vessels, sometimes a vessel in the brain breaks, and a small amount of blood escapes into the brain substance, where it clots. This often produces unconsciousness and paralysis, and may terminate in death. If the amount of blood thus set free be large enough, or if it be in the most vital parts of the brain, it may cause instant death. This bursting of a blood vessel and consequent pressure on the brain causes the disease known as apoplexy.

The clotting of the blood is due to the change of some of its liquid elements into a substance called fibrin. Fibrin consists of innumerable delicate fibrils, so minute that they are seen only with the higher

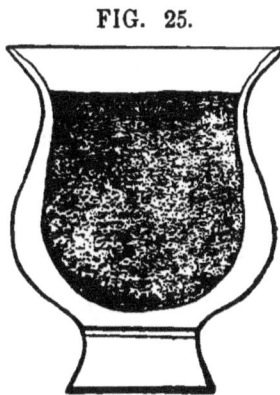

FIG. 25.　　　　　　　FIG. 26.

FIG. 25. A bowl of recently coagulated blood ; the clot is of uniform density.
FIG. 26. The same bowl of blood, a few hours later ; the clot is contracted and floats in the liquid serum.

powers of the microscope. The fibrils are like so many minute threads, which entangle the blood corpuscles and form with them a soft, semi-fluid mass. In a few moments after blood has been exposed to the air, it begins to change to a jelly-like mass. Still later the mass begins to contract, while there escapes from it a clear fluid, called the serum. Later on the central mass becomes quite hard, so that it may be cut with a knife. This central hard mass is known as the clot, and consists of the fibrin and the corpuscles; while the serum represents the other constituents of the blood.

The Blood of the Lower Animals. In many animals the red corpuscles are of the same shape as in man. With the one single exception, the camelidæ, this is true of all the mammalia, — as the horse, sheep, ox, hog, and dog. While the red corpuscles of these animals are of the same shape, yet many of them differ in size. In many of these animals the red corpuscles are so much smaller than those in man that it is sometimes possible to tell whether the corpuscles found in a blood stain are those of man or of some lower animal. To determine this requires the very highest powers of the microscope, very delicate measuring instruments, and also great skill in their use. It is not always possible to tell human blood from the blood of other mammalia, but it is possible to do so in certain cases.

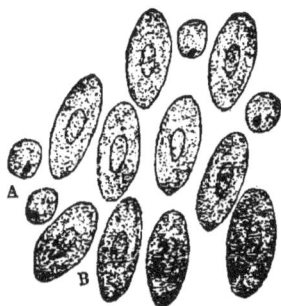

Many of the lower animals have red corpuscles of different shape from those of man. They are also larger and of different structure. The red corpuscles of the blood of birds, fishes, reptiles, frogs, and toads are oval in shape, and much larger than those of

FIG. 27. Frog's blood, highly magnified : (A) the white corpuscles; (B) the nucleated, oval, red, corpuscles.

man. It is comparatively easy to tell these corpuscles from the circular ones, as a glance at Figs. 23 and 27 will show.

Advantage is often taken of these facts, when great crimes have been committed, in order to determine whether a blood stain was caused by human blood or by the blood of one of the lower animals. Thus a knowledge of histology and the use of the microscope often aids in detecting crime and bringing the guilty to punishment.

Fig. 28 illustrates what has been said. The red corpuscles from thirteen different animals are illustrated, showing their relative sizes. Six are from the mammalia, showing a difference in size but not in shape; they are therefore easily told from the others which are oval. The first six have no nuclei, while the nucleus shows prominently in each of the others.

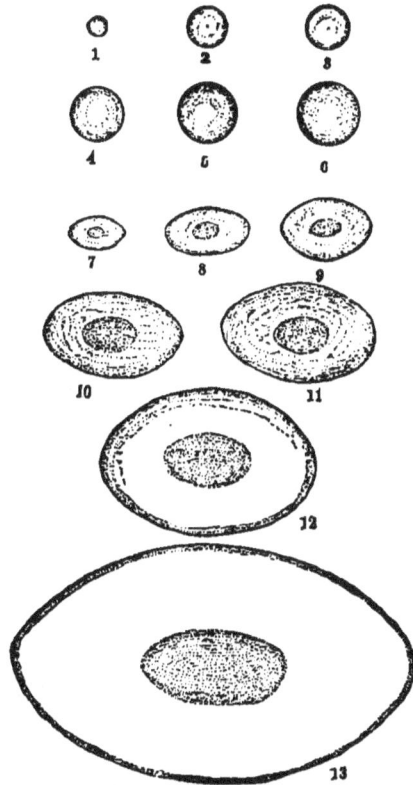

FIG. 28. Showing the relative size of red blood corpuscles of different animals: (1) musk deer ; (2) horse ; (3) mouse; (4) man ; (5) whale ; (6) elephant ; (7) humming bird ; (8) pheasant ; (9) pigeon ; (10) snake ; (11) crocodile ; (12) triton ; (13) proteus.

ALCOHOL AND THE BLOOD.

It is the testimony of eminent surgeons that wounds do not heal so well on those who use alcoholic drinks. The continued use of alcoholic beverages results in such changes in the blood that the fibrin is not so easily formed ; and as the formation of fibrin is necessary for the coagulation of the blood, it follows that bleeding is not so easily checked. Therefore surgical operations are not so successful when performed on those whose systems have been weakened by the long use of alcoholic drinks.

QUESTIONS.

1. What is said about the loss of blood causing weakness and death?
2. What do you understand by the transfusion of blood?
3. Name some parts in which blood is not found.
4. Explain why the blood is a medium of exchange.
5. What is the composition of the blood ?
6. How many kinds of blood corpuscles are there ?
7. Describe the white corpuscles.
8. Describe the red corpuscles.
9. Explain the function of the red corpuscles.
10. Where is arterial blood found? Venous blood?
11. How necessary is oxygen to the body ?
12. How is carbonic acid formed in the body?
13. Give some of the differences between arterial and venous blood.
14. Describe the clotting of blood. What causes it ?
15. Of what does the clot consist? The serum ?
16. What animals have red corpuscles of the same shape as in man ?
17. What animals have red corpuscles of different shape from those of man ?
18. In what way does alcohol affect the blood ?

CHAPTER XI.

THE CIRCULATION.

The Circulatory Apparatus. We have learned that the blood is a great circulating medium by which nourishing material is carried to all parts of the body, and other material brought to organs and tissues to be cast out as of no further use. To bring this about there are four different parts to the circulatory apparatus, — the heart, the arteries, the capillaries, and the veins.

The Heart. The heart is situated in the thoracic cavity, between the lungs. It is completely surrounded by a sac or membranous bag, called the pericardium, the lower part of which rests on the diaphragm. This sac also encloses about two inches of the large blood vessels at the base of the heart. The cells which cover the inside of the pericardium secrete a watery fluid which keeps its inner lining very smooth and enables the heart to move against it without friction. The heart itself is a large, hollow muscle, weighing from ten to twelve ounces, and measuring about five inches in length.

The heart is placed obliquely in the chest. It is conical in shape, with the apex of the cone pointing downward, forward, and to the left. The location of the apex can be easily determined by placing the hand over the left side, and feeling the strokes against the walls of the chest. The upper end or base of the heart is on a level with the cartilage of the third rib.

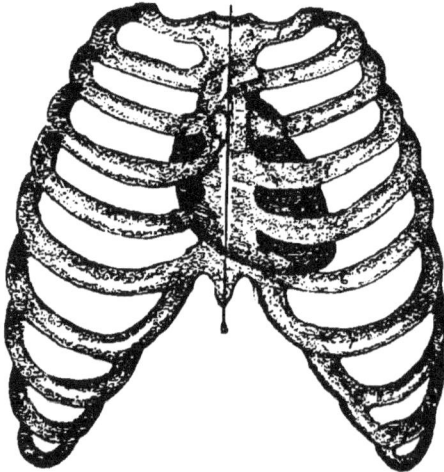

A reference to Fig. 29 shows that the heart is not all on the left side of the body. It extends about three inches to the left of the median line, and an inch and a half to the right of it. The apex of the heart is well over in the left side, and as low down as between the fifth and sixth ribs. The apex has considerable freedom of motion, while the base has but little motion.

FIG. 29. The position of the heart.

The Cavities of the Heart. The heart is divided lengthwise by a firm muscular wall. There is no connection whatever between the two parts thus formed; one side always containing arterial blood, and the other side venous blood. Each side of the heart is again divided by a wall, which, in this case, is not complete; an opening is left so that blood freely passes from one part to the other. This opening is protected by valves which allow the blood to pass only in one direction. There are, therefore, four compartments or cavities in the heart. The two upper cavities are called the auricles, from their fancied resemblance to ears; while the two lower

PLATE IV.

cavities are called the ventricles. Each side of the heart, therefore, consists of an auricle above and a ventricle below. The ventricles have no communication with each other, neither have the auricles; but each auricle has an opening into its own ventricle, protected by valves. All these cavities are lined by a very smooth membrane. The colored plate is drawn from the same specimen as Fig. 30. It represents the right side of the heart as filled with the dark (blue) venous blood, and the left side filled with the bright (red) arterial blood.

A study of Fig. 30 will aid to a better understanding of the above facts. The right side of the heart is represented at 3 and 4, while 6 and 7 represent the left side. The wall between the two is directly beneath the small blood vessel shown on the outside of the heart, to the left

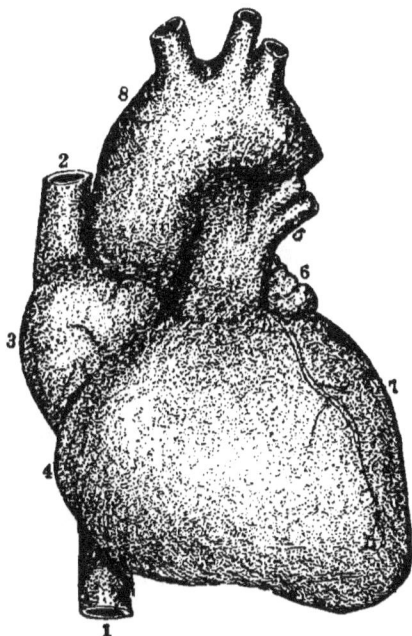

FIG. 30. The heart and the larger vessels at its base or upper part: (1) and (2) veins; (3) right auricle; (4) right ventricle; (5) pulmonary artery; (6) left auricle; (7) left ventricle; (8) aorta.

of the number 7. Several blood vessels course over the heart, giving its muscular tissue proper nourishment. A careful study of each side of this organ will show that the right side is divided into two parts by a cross partition. These parts are represented by the numbers 3 and 4, the former being the right auricle, and the latter the right ventricle. The large vessels, 1 and 2, go to the right auricle, and one large vessel, 5, proceeds from the base of the right ventricle. The left side of the heart is partly hidden from view, still the numbers 6 and 7 show the corresponding left auricle and ventricle. Coming from the left ventricle is the largest artery in the body, 8, the aorta. The vessels entering the left auricle are hidden from view, but they are represented in the following diagram. Fig. 30 is a fairly correct anatomical representation of the heart, with the relative size of its various parts, while Fig. 31 is decidedly a diagram.

The Contractions of the Heart. The auricles always contract together. This contraction is immediately followed by that of the ventricles; these also contract together. When the auricles contract they force their contents into the ventricles; when the ventricles contract the valves between the auricles and ventricles close, so that the blood cannot flow back again. Thus the blood is forced out into the arteries. After the ventricles cease to contract and again relax, the blood is prevented from flowing back into them by means of valves. Therefore, valves are found between the auricles and the ventricles, and between the ventricles and the arteries. Should these valves fail to close and thus to prevent the blood from flowing in the wrong direction, the most serious results might follow. Occasionally the valves become so diseased that they do not completely stop the flow of blood backward, producing a peculiar murmur in the heart sounds, which the physician is able to detect at once. There is great danger from such a condition, as it often produces instant death.

Each ventricle holds from four to six ounces of blood. The earlier investigators placed the amount even below four ounces, but a number of the most recent authorities place it fully as high as six ounces.

The Course of Blood through the Heart. The circulation of the blood through the heart is as follows: from large veins into the right auricle; then through the right ventricle; from this through the lungs; then through the left auricle; and finally through the left ventricle into the large aorta at its base. Briefly: right auricle, right ventricle, lungs, left auricle, left ventricle.

A careful study of the diagram, as represented in Fig. 31, will render this more clear. The blood is returned from the tissues to the heart by numerous veins. These keep uniting, until at last there are only two. The large vein represented at 1 returns the blood from the lower part of the body, while 2 is the vein which returns the blood from the upper part of the body. Both these large veins pour their venous blood into the right auricle, 3. When this becomes filled to its normal limit it contracts and forces the blood through the opening in the direction of the arrowhead, into the right ventricle, 4. When this is filled, it contracts. As soon as it does so, however, the blood is forced back against the half-opened doors, the valves, causing them to close suddenly.

As no blood can return into the auricle, 3, it is forced into the small opening at 5, which is the beginning of the pulmonary artery. When the right ventricle again relaxes, the blood from the pulmonary artery cannot flow back into it, owing to the closure of the valves, represented by V immediately below the figure 5. The dark blood then flows through the lungs, giving off its carbonic acid and receiving a fresh supply of oxygen. It finally passes through the lungs, entering the four

pulmonary veins, at 6, as bright arterial blood; these veins bring the blood to the left auricle, 7. When filled the auricle contracts, forcing its contents into the left ventricle, at 8. When the left ventricle contracts, the valves between it and the left auricle close, in the same manner and at the same time as those on the opposite side. The blood is then forced out into the large aorta, 9. From this artery the blood cannot return to the heart, for

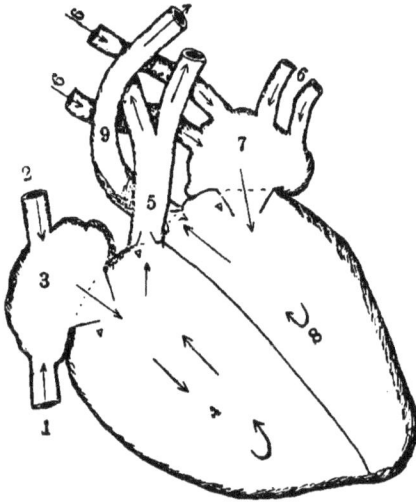

FIG. 31. A diagram illustrating the flow of blood through the heart: (1) and (2) veins; (3) right auricle; (4) right ventricle; (5) pulmonary artery; (6) pulmonary veins; (7) left auricle; (8) left ventricle; (9) aorta.

valves guard the opening here also. On through the arteries the blood flows until the finest capillaries are reached. Here the proper nourishment is given off to the tissues, and some of the worn-out material received in its place. The blood, now dark and impure, returns through the veins to the right side of the heart, only to start again on another journey just such as we have described.

The Left Ventricle Stronger than the Right. The left ventricle has to contract with force sufficient to send the blood to the most distant parts of the body ; while the right ventricle has to send the blood only to the lungs, which are but a short distance from it. Therefore, we should expect to find the left side of the heart stronger ; that is, its muscular walls should be thicker. In this case we are not disappointed, for a cross section of

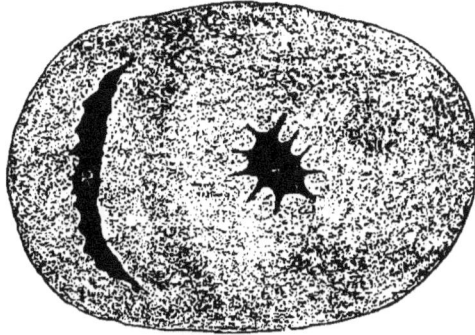

Fig. 32. A cross section of the ventricles of the heart.

the heart through the ventricles appears as illustrated in Fig. 32. This shows the relative thickness of the walls of the ventricles in their contracted state, and also the difference in the shape of the ventricles. To the right of the figure is the nearly circular cavity of the left ventricle, surrounded by thick walls. To the left is the flattened cavity of the right ventricle, with much thinner walls.

The Pulsations of the Heart. The beat of the heart is involuntary and subject to much variation. The average frequency of the pulse in man is about 70 per minute ; while in woman it is about ten more each minute. This average may be greatly increased for a short time by many circumstances, but if the increase be long continued it denotes some disturbance of the system. An excessive rapidity of the pulse, continuously beating 150 or 160 a minute, indicates great danger ; while a very slow pulse, much below the normal, also indicates serious trouble. Some persons naturally have a pulse much above or even below the average. It is said that the pulse of Napoleon I. was but 40 per minute. The pulse varies greatly with the age. At birth it is about 130. It gradually falls until at three years of age it is about 100; at fourteen, about 80; and at twenty-one, about the average for the adult.

Muscular exercise accelerates the pulse. The muscular effort of standing makes a normal difference of ten beats per minute more than when lying down. Active exercise, as running and jumping, increases the pulse. Excitement, as joy and anger, increases the number of beats; while sorrow and depression of spirits may cause the number to be reduced. Any sudden excitement, as fright, will cause the heart to beat violently, so that it is felt to strike with much force against the walls of the chest. Excessive fear, joy, or grief, may have an effect on the nervous system powerful enough to cause the heart's action to cease, producing instant death. Taking the seasons through, the heart beats faster in summer than in winter.

The Heart Works and Rests. The heart does an immense amount of work. At the rate of seventy beats per minute, there are a hundred thousand contractions daily. The labor expended by the heart each day is equal to a force required to lift 120 tons a foot from the ground.

If the heart does such an amount of work, it must have rest. We find there is a period of time when it is completely at rest. The auricles contract together; immediately after, the ventricles contract, also; and following their contraction there is a period of complete rest, after which the auricles contract again. Brief as this period is, it yet represents about one fourth of the time of a whole beat. From this it is seen that the sum of all these brief periods for a whole day is not less than six hours.

The Sounds of the Heart. If the ear be placed over the heart two distinct sounds are heard, one immediately following the other. After a moment of silence they are repeated. It is noticed that these two sounds correspond with each beat of the heart. The first sound is comparatively long and dull ; the second, almost immediately following it, is sharper, shorter, and more distinct. The sounds are likened to those produced by pronouncing the words *tŭbb, dŭp.* The whole time of the pulsation of the heart may be divided into four parts ; the first sound, two parts ; the second sound, one part ; and the period of rest, one part.

The first sound is caused by the sudden closure of the valves which are between the auricles and the ventricles ; together with the sound caused by the powerful contraction of the muscular walls of the ventricles. The first sound is, therefore, a valve sound and a muscle sound. The second sound is caused by the sudden closure of the valves which are at the beginning of the pulmonary artery and the aorta. It is, therefore, entirely a valve sound.

The Pulse. During life the arteries are always full of blood, but as their walls are elastic they can be distended so that at times they may contain more blood than at others. Each contraction of the heart suddenly forces a quantity of blood into these elastic tubes, distending their walls. When the heart relaxes, the over-distended arteries would force the blood back into it were it not for the closure of the valves; these fully prevent any backward flow. As the blood cannot go in a backward direction, it is pushed forward under the pressure of the elastic walls of the arteries. Thus the arteries relieve themselves of the excess of blood, so that their walls are not so fully distended. But no sooner have the arteries returned to their former size than they are again expanded by another contraction of the heart. This series of expansions of the arteries give rise to the pulse, which is present in all arteries. Therefore each arterial expansion, or each pulse, represents a contraction of the ventricles. The pulse thus becomes a guide for ascertaining the frequency and regularity of the heart's action, and the condition of the general circulation.

As nearly all the arteries are deep-seated, only those few near the surface are used to study the pulse. The radial artery at the wrist is usually chosen for this purpose, although the pulse may be felt on the temple, the neck, and other places. It is possible to see the pulsations of an artery with the unaided eye, such pulsations showing at times on the temple, on the neck, or at the wrist.

Arteries. The arteries are the vessels which carry the blood from the heart to the various parts of the body. The word "artery" is derived from two Greek words which signify "receptacle of air;" for the ancients believed that these vessels contained air in the living body. This belief was probably founded on the fact that the arteries are usually found empty in the dead body. This condition exists because after the heart ceases to beat, the elastic walls of the arteries contract with sufficient force to push all their contents forward into the capillaries and veins.

The arteries are cylindrical, firm, and elastic canals. Their lining membrane is a continuation of the membrane lining the heart. Their walls are composed largely of elastic and muscular tissue. The strong elastic tissue allows the artery to expand without danger of bursting. The muscular tissue gives contractile power to the artery, so that it can accommodate itself to the amount of blood it contains. The muscle is of the involuntary variety, and the cells are arranged around the tube, so that their contraction will diminish the size of the canal. In the smaller arteries this power is of great use. The muscle coat is very prominent there, and its degree of contraction determines the size of the vessel. As the contraction of the muscle is under the control of the nervous system, so there are many circumstances which will affect the size of the artery.

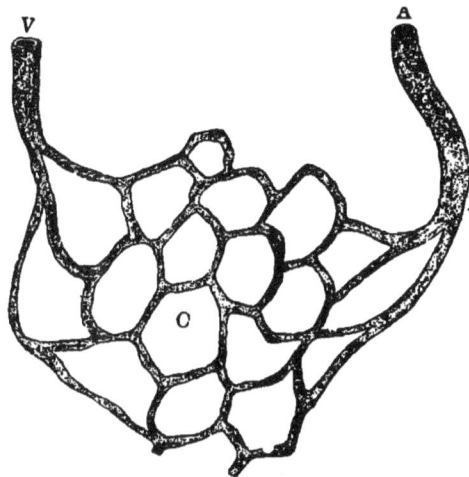

FIG. 33. A, a small artery ; c, capillaries ; v, a small vein.

The muscular and elastic coats cause the artery to retain the circular shape when it is cut across, so that it resembles a tube. After repeated divisions of the arteries they become

8

very small, so that they can only be seen with the highest powers of the microscope. Even the muscular and elastic coats disappear, and only the thin lining membrane is left.

The Capillaries. When the artery has become so small that it has only the thin membrane for its wall, it is called a capillary. The capillaries are the most minute blood vessels which penetrate the organs and tissues of the body. They bring the blood in very close contact with the cells of the tissues; for between the living tissues and the blood in a capillary, there is but the thinnest membrane. Many of the capillaries are so small that when the blood flows through them the corpuscles have to pass in single file, as there is not room enough for them to flow side by side.

The Veins. Soon the capillaries unite with each other to make larger vessels, and their walls become thicker. The vessels are now called veins. The veins convey the blood from the capillaries toward the heart. The smaller veins unite to make larger ones, until at last the large veins are formed which empty into the right auricle of the heart. The walls of the veins are much thinner and less elastic than the walls of the arteries. If a vein be cut across, it will collapse and appear flat, because there is but little muscular and elastic tissue in its walls.

Nearly all the veins have valves to prevent the backward flow of the blood. This is illustrated in the accompanying figures. It is evident that if the blood flows in the direction of the arrow-head, in Fig. 34, the valves will remain open, but if it should attempt to flow in the opposite direction, as in Fig. 35, the valves would close and completely shut off the passage.

Rapidity of the Circulation. The blood nearest the heart, in the aorta, flows the most rapidly ; for all the force of the heart's contraction makes itself felt here. As the arteries divide, the stream becomes less rapid until

FIG. 34.   FIG. 35.

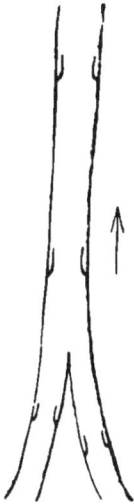
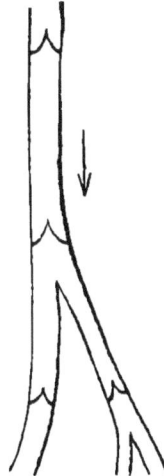

FIG. 34.   The valves of a vein, open.
FIG. 35.   The valves of a vein, closed.

in the capillaries it is much slower. It is estimated that the blood in the aorta flows five hundred times faster than it does in the capillaries. In the veins the flow is faster than in the capillaries, but it does not equal the speed acquired in the arteries. In the large arteries the blood flows at the rate of about a foot per second. A quantity of blood can leave the heart, make a complete circuit of the circulation, and reach its starting place again in less than half a minute.

The General Circulation. A study of Fig. 36 will give an idea of the manner of distribution of the principal blood vessels. In this figure, N represents the neck ; D,

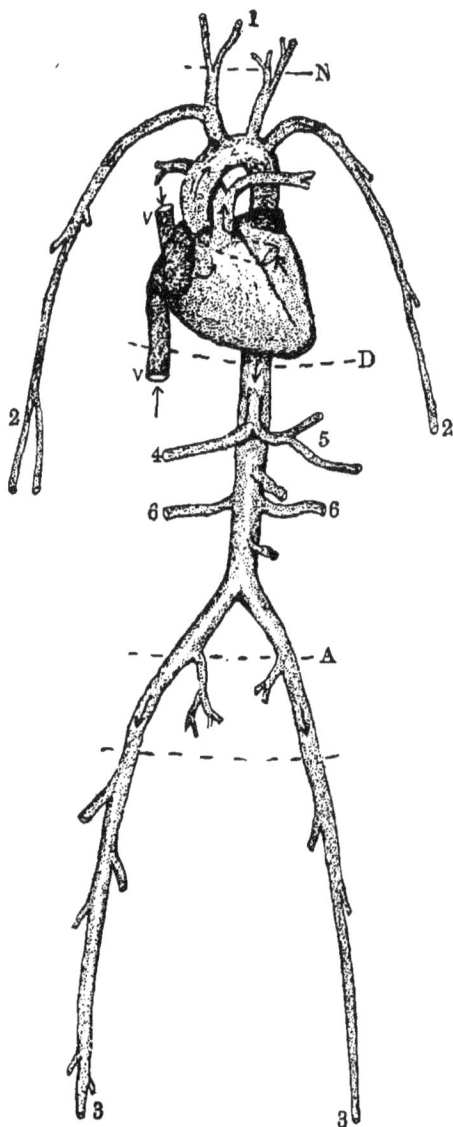

the diaphragm ; and A, the lower border of the abdominal cavity. Let us follow the course of the largest artery in the body, the aorta, which is nearly an inch in diameter. It is represented in Fig. 30 at 8, and in Fig. 31 at 9. In Fig. 36, its beginning is shown by the dotted lines, giving its origin from the left ventricle. It passes upward a short distance, then arches backward to the left, and so descends by the side of the vertebral column. It passes down through an opening in the diaphragm, D, and continues down the abdominal cavity, until at its lower part it divides into two branches, which supply the lower extremities with blood.

FIG. 36. A diagram, showing the general plan of the circulation : N, the neck ; D, the diaphragm ; A, the lower border of the abdominal cavity ; V, veins returning the blood to the right side of the heart. All other vessels represent arteries, which carry blood to the following parts : (1) to the head ; (2) to the arms ; (3) to the legs ; (4) to the liver ; (5) to the stomach, pancreas, and spleen ; and (6) to the kidneys.

From the arch of the aorta, branches arise which supply the head with blood, 1, and which send large vessels to the upper extremities, 2. The rather irregular manner in which these branches are given off from the arch is here correctly represented. The branches going to the left side of the head and the left upper extremity arise separately, while, on the right side, they arise from the aorta as one branch and afterwards divide.

In the abdominal cavity, just below the diaphragm, the aorta gives off a large branch which divides into three branches; these supply the liver, 4, and the stomach, pancreas, and spleen, 5. Lower down is a single branch, which goes to the small intestine. Still lower are the two arteries, 6, which go to the kidneys. Below these is a branch which supplies blood to the large intestine. Thus the aorta is the great central artery of the body. It is deeply seated in the thoracic and abdominal cavities, and is well protected from injury. The veins are placed side by side with the arteries.

Aids to the Circulation. Exercise is one of the greatest aids to the circulation. It not only causes the heart to beat faster and thus hastens the flow of blood, but it is a direct aid to the movement of blood in the veins. When a vein is filled, the blood cannot flow backward toward the capillaries on account of the valves; therefore, if the muscles be made to contract and thereby press upon the veins, the blood will be pushed onward. When the muscle relaxes, the vein is again filled with blood coming from the capillaries. More muscular exercise will again hasten on the blood to the heart. The surface of the body should be kept warm, because cold contracts the blood vessels, and thus interferes with a free circulation. Very tight clothing is also a hindrance to the free flow of the blood. The clothing may fit closely and yet not so tightly that the flow of the blood in the veins is impeded.

THE EFFECTS OF ALCOHOL AND TOBACCO.

Alcohol and the Heart. The heart is an involuntary muscle ; it keeps steadily at work, whether we " will " to have it or not. While it is not under the control of the will, yet, like all other muscles, it is controlled by nerves. Experiments have proved that there are nerves which act as brakes to the heart, holding it in check and regulating its beat.

In a later chapter will be stated more fully how alcohol tends to benumb, or partially paralyze, the nerves. So alcohol tends to paralyze the nerves that hold the heart in check. As a result, the heart beats faster. One result of this extra work is that the muscles of the heart become thickened and the cavities enlarged. The heart is not in repose long enough to have its overworked tissues fully nourished, and so a process of degeneration begins. Particles of fat, or oil, take the place of the muscle. At first, this is very slight; but gradually a change occurs, until much of the worn-out muscle disappears and fatty tissue takes its place.

The heart now becomes very weak, hardly able to force the blood through its proper channels. The circulation is poor, the extremities are swollen, and there is difficulty in breathing. The physicians call this trouble "fatty degeneration of the heart." The heart becomes weaker and weaker, until, suddenly, it is unable to do its work longer, and death occurs. This is no idle story. The effect of alcoholic beverages is so marked that fatty degeneration of the heart is often called, by physicians, the "whiskey heart," or the "beer-drinker's heart."

It is well to remember that the strongest drinks are not necessarily required to produce these changes. The extra amount of flesh found in those who use ale and beer freely is of this fatty nature; and a fatty heart is no more than we would expect to find in such a fleshy body. Physiologically, alcohol over-works the heart; anatomically, alcohol changes its structure. Alcohol never imparts life-giving power, while it often causes death.

Alcohol and the Arteries. The same changes already mentioned, as occurring in the muscular tissue of the heart, may also take place in the muscular tissue of the walls of the arteries. The cells become weakened by the deposit of fat within them, and thus the whole wall loses some of its strength. This makes the walls more liable to rupture, and is one of the predisposing causes of apoplexy.

Alcohol and the Smaller Vessels. The walls of the smaller arteries consist largely of muscular tissue. This tissue is continuously kept in a partially contracted condition, in order that the walls may be more firm. The muscular tissue is under the control of minute nerves, called the vaso-motor nerves. These nerves are capable of making the muscle contract firmly, thereby diminishing the size of the vessel; while if they cease to act, from any cause, the muscle relaxes and the size of the vessel is greatly increased.

The small arteries are said to have "tone" when these nerves exert their power, and keep the muscle in its usual condition of moderate contraction. The walls of the smaller arteries, therefore, are constantly in a state of moderate contraction. The effect of alcohol is to partially paralyze these vaso-motor nerves, so that they cease to exert their full power on the muscular tissue. If a small dose of an alcoholic beverage be taken, the effect may be slight and only temporary, as illustrated in the flushing of the face; if not repeated, recovery may be complete; but a continuation of such doses causes the paralysis to become more permanent, and the nerves to lose their power of controlling the size of the blood vessels.

Not only the red nose and the red eyes of the confirmed drinker, but the reddened face and distended capillaries often seen in the moderate drinker, are indications of the paralysis of the vaso-motor nerves. The "tone" has disappeared from the walls of the blood vessels; the muscular tissue is becoming or has become permanently relaxed, and the vessels are constantly in a distended condition. But the nose, eyes, and other portions of the face are not the only places where this congestion occurs. It exists in the mucous membranes of the body to a large extent; while many of the organs and tissues are also in this chronic congested condition.

If such a condition of things were to arise from any other cause the individual would be greatly worried and would seek medical advice at once. But he either does not realize that the alcohol is the cause of all the aches and pains he feels, or cares too much for it to give it up, and thus he continues his downward course. We insist that the man who is under the influence of powerful poison is a sick man, no matter whether he be intoxicated, or simply indulges in moderate doses. In either case his system is affected, and he should be under the careful attention of a competent physician.

Such men need medical advice and restorative remedies far more than the vast majority of those who are regular patrons of the physician's office. The burning of the stomach, the darting neuralgic pains, the headaches, all these and many other ailments indicate the tremendous damaging power of this poison. Nothing short of a strong resolve to use no more alcoholic liquors will answer. To attempt to stop gradually is a very poor plan; but even a firm resolve, the breaking away from old associates, and the forming of new, may not prove sufficient. The old appetite remains. It is here that the competent physician can accomplish much. He can aid nature in her efforts to restore the body to health. He can aid in building up the tissues, in correcting the congested and inflamed organs, and thus banish from the system an appetite as unnatural as it is injurious.

But how can all this be prevented? How can we all escape the strong power of alcohol? Simply by refusing to take the first glass of any kind of alcoholic drink.

Tobacco and the Heart. Tobacco affects the heart largely through the action of the nicotine on the nervous system. A prolonged use of tobacco frequently gives rise to a particular affection known as the "tobacco heart." The author has seen a strikingly large number of these cases in young men between fifteen and twenty years of age. The heart is irregular in its action and sometimes beats with great force. This is often accompanied with a sensation of weakness or great anxiety. There are occasional attacks of dizziness, shortness of ' breath, nausea, and vomiting. At times there is intense pain in the region of the heart. Tobacco will not make so serious changes in the structure of the heart as have been described under alcohol, yet it is capable of doing immense harm.

Within a few years there have been well-known instances among our most prominent statesmen, where death was instantaneous, as a result of disease of the heart from the long-continued use of tobacco. With such a history, tobacco should no longer be regarded as a luxury, but rather as a slow poison capable of accomplishing an enormous amount of damage.

QUESTIONS.

1. Name the four parts of the circulatory apparatus.
2. Describe the heart, and its position in the chest.
3. Describe the cavities of the heart and the valves between them.
4. Give the course of the blood through the heart.
5. Which ventricle is the stronger?   Why?
6. How many sounds of the heart?   Describe them.
7. What causes the pulse?
8. What is the structure of the walls of the arteries?
9. How do the veins differ from the arteries?
10. How rapid is the circulation?
11. Give some of the effects of alcohol on the smaller vessels.
12. How does tobacco affect the heart?

SUGGESTIONS TO TEACHERS.

1. **The Heart.** Procure the heart of a calf or sheep at the market. Preserve the large vessels at its base. Wash in water and wipe dry. Call attention to the shape; to the auricles; and to the ventricles. Hold the heart obliquely, as in Figs. 29 and 30, thus showing its position in the body.

2. **Circulation through the Heart.** This can be illustrated by pointing to the parts in order, thus : right auricle, right ventricle, left auricle, left ventricle, according to Fig. 31. The right side of the heart can be told from the left by remembering that the left ventricle projects around to form the point of the heart and a small part of the anterior surface. The walls of the right side are also thinner.

3. **The Arteries.** Notice the size and the thick walls of the aorta. It keeps open, as a circular tube. Lift the heart by the aorta, and notice the elasticity of the artery.

4. **The Veins.** These will be collapsed, with thinner walls, which are not elastic like the aorta.

5. **Walls of the Ventricles.** Cut the heart open transversely about half way back from the apex, to illustrate Fig. 32. The firm partition between the ventricles, the relative thickness of the walls, and the smooth lining-membrane, will be brought to view.

6. **Valves.** Cut away the ventricles close to the auricles. Notice the thin membranes which are between the auricles and ventricles ; these are the valves.

7. **Pericardium.** By making arrangements at the market, a heart may be procured, surrounded by its sac. This is the pericardium. Cut it open with scissors. A small amount of fluid may escape. Notice how smooth is the lining of this membrane.

CHAPTER XII.

RESPIRATION.

The Nasal Cavities. The nostrils are the proper channels through which the air should reach the lungs. The
nose has at least three important functions to perform
in connection with respiration : these are to warm, to
moisten, and to filter the inspired air.

The first of these is evidently very important; for, if
the cold air of winter should be brought directly in contact with the tissues of the throat and larynx, inflammation would be likely to follow, causing sore throat,
hoarseness, and loss of voice. The tissues of the nasal
cavities are so well supplied with blood that they
are capable of warming the air as it passes over them,
until its temperature more nearly equals that of the
body.

The second function is likewise important: there is
at least a pint of serum secreted every twenty-four hours
by the mucous membrane lining the nose : the inspired
air passes through the nose, takes up this moisture, and
becomes saturated with it. That the inspired air takes
moisture from the tissues is easily proven by breathing
through the mouth for a short time. The throat soon
becomes dry, and swallowing is difficult. The cells covering much of the lining membrane of the nose are of

the ciliated variety, as represented at 4, in Plate I. The cilia catch the particles of matter found in dust and smoke, and in the ordinary inspired air. In this way the nose acts as a filter.

Mouth Breathing When breathed through the mouth, the air is but little warmed, is only slightly moistened, and is not filtered. Mouth breathing brings the air in contact with the larynx, trachea, and bronchial tubes, scarcely changed. It is still cold, dry, irritating, and, as a result, more or less inflammation is produced. Inflammation of the throat, enlarged tonsils, chronic hoarseness, and coughs are some of the affections which result from the pernicious habit of breathing through the mouth. Nature intended we should breathe through the nose, and a number of evils will result if we fail in so doing. If it be impossible to get air through the nose, a physician should be consulted that the difficulty may be removed. Early attention to these conditions would do much to prevent the catarrhal affections so prevalent in this country.

The Larynx. After the inspired air has passed through the nose, it enters the upper part of the pharynx. From here it passes down the throat, until opposite the base of the tongue, where it reaches the larynx. The larynx is situated at the upper and front part of the neck. It contains the parts necessary for the production of the voice. The expansion on the front of the larynx, so much more prominent in men than in women, is commonly known as Adam's apple. The larynx is composed of cartilages, lined with a mucous membrane. About the middle of its interior are two strong bands of elastic tissue, called the vocal cords. They extend

from the front to the back of the laryngeal cavity. The
space between them, through which the air passes, is

FIG. 37. FIG. 38.

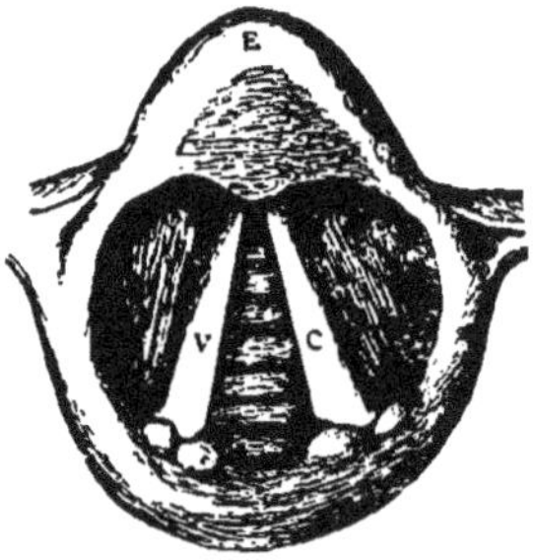

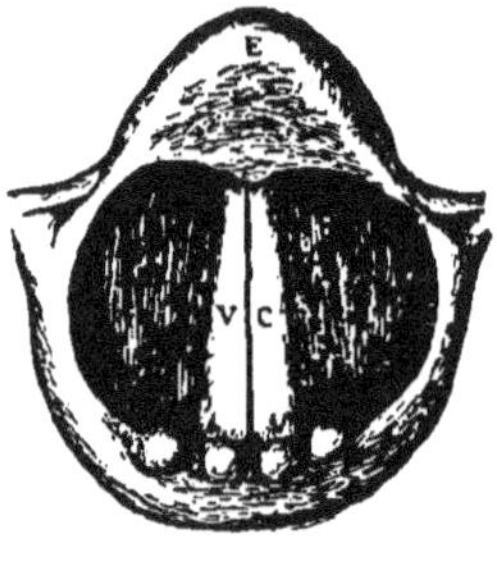

FIG. 37. The position of the vocal cords during inspiration: the
rings of the trachea are seen between the vocal cords.
FIG. 38. The position of the vocal cords when uttering a high note:
v c, vocal cords; E, epiglottis.

called the glottis : this opening varies in size according

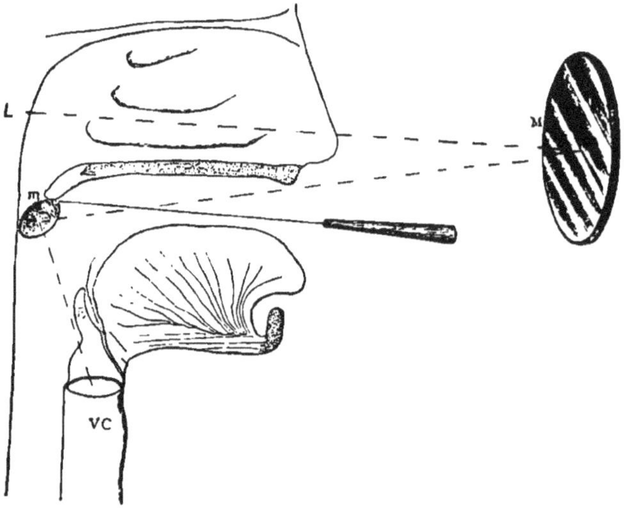

FIG. 39. A diagram illustrating how the vocal cords are seen, as in
Figs. 37, 38 : M, a circular mirror held, by a band, to the forehead of the
operator; L, a lamp, placed at the side of the patient's head, to throw
light on the large mirror, M ; m, a small mirror held in the back of the
throat of the patient : E, epiglottis ; v c, location of the vocal cords.

to the tension of the vocal cords. This is well illustrated by referring to Figs. 37 and 38. During inspiration the vocal cords are quiet, and the opening between them is large, as in Fig. 37; but when sounds are produced the vocal cords come together and the glottis is narrowed. The higher the notes, so much the tighter will the vocal cords be drawn, and so much the narrower will be the glottis, as in Fig. 38. Figs. 37 and 38 are representations of the interior of the larynx, as seen with the laryngoscope.

The laryngoscope consists of a small mirror attached to a long handle. To use it, the mouth is opened, the tongue drawn forward, and the mirror introduced as shown in Fig. 39. Direct sunlight, or artificial light, is reflected from a large

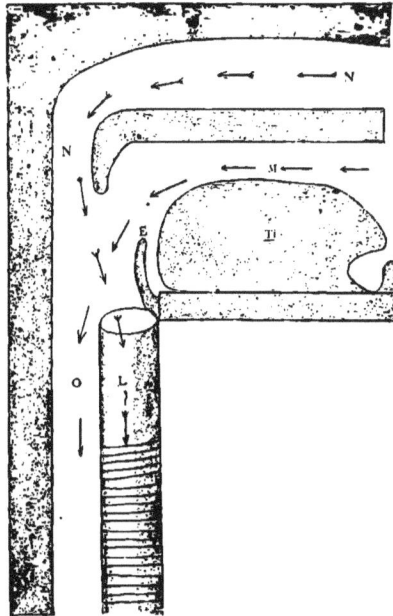

FIG. 40. A diagram illustrating the position and use of the epiglottis : N, nasal passages ; M, mouth ; O, œsophagus ; L, larynx ; T, tongue ; the feathered arrows represent the passage for air ; the plain arrows, the passage for food.

mirror to the smaller one which is in the back part of the throat. The small mirror reflects the light down the larynx so that its interior becomes brilliantly illuminated. A picture of the larynx is thus formed in the mirror, to which the observer directs his eye.

The Epiglottis. The entrance to the larynx is protected by a valve, or lid, called the epiglottis. During respiration the epiglottis is directed upward, so that the larynx is open ; but during the act of swallowing, the epiglottis shuts tightly down over the larynx, preventing the entrance of any solid or liquid. Occasionally, however, a particle of food " goes the wrong way " and slips into the larynx, when a violent cough is necessary for its removal. Fig. 39 gives a correct representation of the parts under discussion. Fig. 40 is a diagram, illustrating more clearly the location and function of the epiglottis.

The Trachea. The trachea, commonly known as the windpipe, consists of a number of rings of cartilage. These rings are not quite complete at the back of the trachea ; however, the tube is completed by a thin membrane. The rings are placed one over the other, separated only by a narrow membrane. They keep the trachea from collapsing ; thus always insuring a free passage for the air. The tube is lined, its whole length, with a mucous membrane. About opposite the upper part of the sternum, the trachea divides into two branches, called the bronchi, one branch going to each lung.

The Bronchi and the Air Cells. After entering the lungs, the bronchi divide again and again, until they are very minute in size. They are everywhere lined with a mucous membrane. A study of Fig. 41 will make these facts more simple. At 1, is the epiglottis, standing guard over the entrance to the air passages below ; at 2, is the larynx, or voice box ; at 3, is the trachea with its rings of cartilage ; at 4, is the right lung, so drawn that the bronchial tubes can be seen within it.

After these have become very small, from their repeated divisions, they terminate in a collection of minute

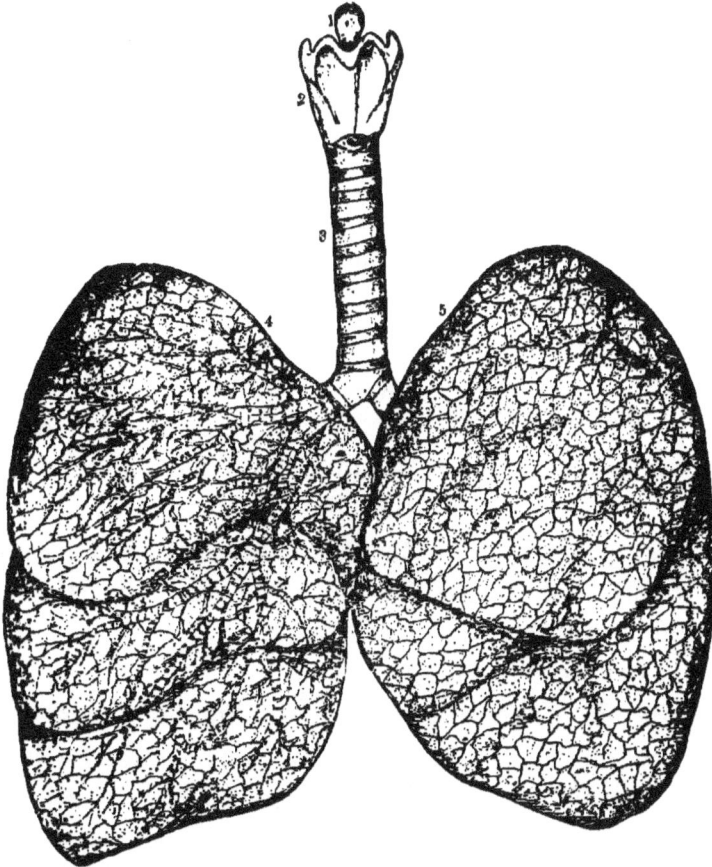

FIG. 41. The respiratory apparatus: (1) the epiglottis ; (2) the larynx ; (3) the trachea ; (4) the right lung ; (5) the left lung.

sacs, called air cells. The walls of these air cells are very thin and highly elastic. They can be distended by slight force, and when the force is removed, they can at once resume their former size. Fig. 42 shows a terminal bronchial tube with its air cells. If we bear in mind that these air cells have elastic walls, it is easy to imagine how they could be inflated, like so many rubber sacs, by forcing air into the tube, at 1. This is practically about what occurs during an ordinary respiration.

The inner surface of these air cells is exposed to the air which enters the lungs. The amount of surface thus exposed is very great, being estimated to be at least fourteen hundred square feet. Surrounding the walls of the air cells is a dense network of capillary blood vessels. Thus the blood itself is separated from the air only by an extremely thin membrane. It is here, in the air cells, that the changes occur which transform the dark venous blood into the bright arterial blood.

FIG. 42. (1) the end of a small bronchial tube ; (2) air cells ; (3) some of the air cells cut open, showing free passage to them from the bronchial tubes.

The Pleura. The pleura is a double membrane, covering the inside of the thoracic cavity and the lungs. One membrane is closely fastened to the inner walls of the chest, while the other covers the surface of the lungs. The space between these membranes is called the pleural cavity. The pleura secretes a fluid, so that its two surfaces may move against each other easily and without friction, as they do in ordinary breathing. An inflammation of this membrane is called pleurisy ; it is extremely painful, because each time a breath is taken the lungs expand, causing the inflamed membrane covering the lungs to move against another inflamed membrane lining the walls of the chest.

The Lungs. The two lungs are situated in the thoracic cavity, one in either side of the chest. Owing to the amount of elastic tissue in the air cells, the lungs have great elasticity. When removed from the body they appear completely collapsed; still they float in water on account of the air yet remaining in the air cells. When in this collapsed condition, if a tube be placed in the trachea the lungs may be inflated by blowing into the tube or by forcing air in with a pair of bellows. After the inflation, it is only necessary to remove the tube or allow the air to escape through it, when the lungs will immediately collapse again. Thus it is easily proved that if some force be applied to send air into the lungs, the elastic tissue in the air cells will stretch like rubber; and that as soon as the force is removed the elastic tissue will return to its former condition.

Why Air enters the Lungs. The mechanism of respiration is not unlike that of a pair of bellows. When the handles are raised the inside of the bellows is made larger, and the air rushes in to fill the extra space.

The chest is a tight box, with only one opening, and that at the top, — the larynx. If this box be suddenly enlarged the air will rush in through the opening; this is called inspiration. When the box ceases to enlarge, no more air enters. Immediately all the parts that were under a tension to enlarge the box return to their former condition. Thus the box is made smaller and the air rushes out of the opening at the top; this is called expiration. From this we conclude that air enters the lungs because the chest is made larger; and that it leaves

the lungs because the chest is made smaller. Before
discussing this subject more in detail, it will be well to
fix in the mind the shape of the thoracic cavity, its contents, and the relation of certain organs to each other.

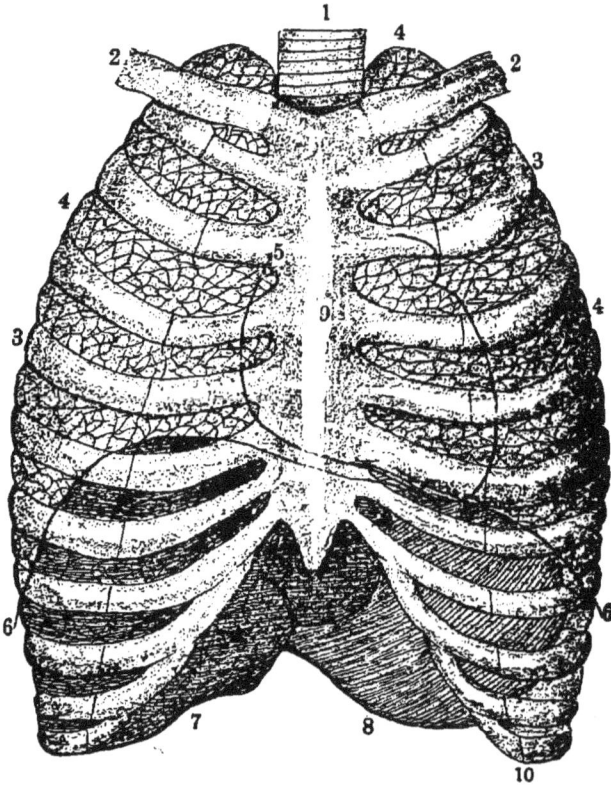

Fig. 43. The position of the lungs, and their relation to certain organs :
(1) the trachea ; (2) the clavicle ; (3) the ribs ; (4) the lungs; (5) a dark
curved line, showing the position of the heart ; (6) the diaphragm, extending in a curved direction from one figure to the other ; (7) the liver;
(8) the stomach ; (9) the sternum.

A reference to Fig. 43 will illustrate these points. The
lungs rise above the collar bone, 2, to form the apex of
the chest : below, they rest upon the curved diaphragm,
6, which divides the thoracic from the abdominal cavity.

The relative position of the heart is shown by the curved line, 5. Pressing up against the under surface of the diaphragm, on the right side is the liver, 7, while on the left side is the stomach, 8. By glancing at this figure it is easily understood that if the lower ribs be brought tightly together from any cause, as by tight lacing, the stomach and liver will be forced up against the diaphragm. As a result of this it would be extremely difficult for the diaphragm to become straighter or lower, or, as illustrated in the figure, for the line 6 to become shorter. The application of this fact will be seen later.

Inspiration. The chest is enlarged first, by the raising of the ribs. By the contraction of certain muscles the ribs are moved upward and outward; this enlarges the chest anteriorly and laterally. That the chest is enlarged between the sternum

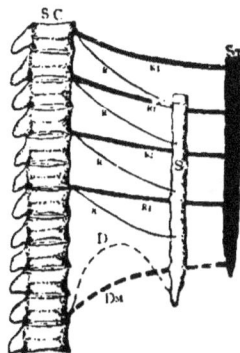

FIG. 44. A diagram illustrating how the thoracic cavity is enlarged during inspiration: S C, spinal column; S T, R I, and D M, illustrate the position of the sternum, ribs, and diaphragm, during inspiration; S, R, and D, their position during expiration.

and the spinal column is readily seen by referring to Fig. 44. S C represents the spinal column, and S, the sternum. After expiration, when the chest is at rest, it is noticed that the ends of the ribs attached to the sternum, S, are much lower than the ends attached to the spinal column; this fact is also illustrated in Figs. 29 and 43. Now in Fig. 44 it is evident that if the ribs, R, are raised by any force, they will carry the sternum with them to the point S T. It is also evident that the distance between S C and S T is much greater than that between S C and S. Therefore we conclude that raising the ribs increases the size of the chest from front to back. Because the ribs are curved, the chest is enlarged from side to side when they are raised.

Second, the chest is enlarged by the contraction of the diaphragm. Attention has already been called to this thin muscle. When it relaxes it is in the form of a vaulted partition, with its rounded portion rising into the cavity of the chest, as represented in Fig. 43, and at D, Fig. 44. When the diaphragm contracts, it shortens, assuming more nearly a straight line, as in D M, Fig. 44; thus the convexity becomes greatly diminished. The contraction of the diaphragm, therefore, makes it descend toward the abdomen; this must enlarge the thoracic cavity from above downwards. Thus we conclude that in ordinary, quiet breathing we do not draw the air into the lungs; the air rushes in, without aid, to fill the chest, which is made larger by the elevation of the ribs and the contraction, or lowering, of the diaphragm.

Expiration. Ordinary expiration occurs without the aid of muscles. By the relaxation of the muscles of the chest, the ribs fall back to their natural places. The relaxation of the diaphragm causes it to protrude again into the chest, aided by the pressure of the abdominal organs from beneath pushing up against it. The distended elastic tissue of the air cells now seeks to return to its natural condition, thus diminishing the size of the air cells, and, consequently the size of the whole lungs; the thoracic walls keep in contact with the diminishing lungs, and thus the air rushes out of the larynx.

Respiration. Respiration is the function by which oxygen is introduced into the body, and carbonic acid removed from it.

Ordinary breathing is involuntary. We breathe when we are not thinking of it, and breathe as regularly when asleep as when awake. But respiration is also voluntary, for it is possible to breathe more slowly or more rapidly than usual, for a short time. By taking advantage of this fact the respiratory muscles can be greatly strengthened. Like other muscles they can be developed and strengthened by exercise. Singing and speaking call for a full exercise of these muscles, and when properly employed they are very beneficial.

The imitations of crying, laughing, and the like are voluntary respiratory movements, principally spasmodic contractions of the diaphragm. Even these movements may be involuntary; the spasmodic action of the diaphragm may go beyond control, so that it is impossible to cease laughing or crying.

Number of Respirations. The number of respirations should be about one to every four beats of the pulse. As the average pulse of the male adult is about 70, so the number of respirations should be about 18. But this is influenced by many conditions; as, the size of the lungs, the condition of the air, exercise, singing, speaking, and many other circumstances. The number cannot, however, be lowered for any considerable length of time; the demands of the system for more oxygen, and for an escape of the carbonic acid are so great that it is impossible to resist them.

Sounds of the Chest. When the air rushes in and out of the lungs, peculiar sounds are produced. These are easily changed by unhealthy conditions of the lung tissue; the air may not enter a portion of the lungs; it may enter a cavity; or it may pass over a membrane either too dry or too moist. Any such departures from health cause a change in the normal sounds. A careful study of these sounds enables the physician to determine the healthy or unhealthy condition of the lungs. By their aid, he can understand the nature of the disease, and how far it has progressed.

The Inspired Air. Each inspiration brings about twenty cubic inches, or two thirds of a pint of air into the lungs. This air only extends down the lungs a short distance, probably not much below the larger bronchial tubes. If the air should remain here it would be of little use. But the law of the diffusion of gases is such that the gases — the oxygen and the carbonic acid — which are in the bronchial tubes and the air cells freely and rapidly intermingle. The inspired air brings oxygen to the red corpuscles of the blood.

The Expired Air. If the expired air be collected and examined, it will be found to differ from the inspired air in the following particulars : 1. It has lost oxygen. The expired air contains nearly five per cent less oxygen than the atmospheric air. 2. It has gained carbonic acid. The expired air contains nearly a hundred times more carbonic acid than the atmospheric air. This gas represents one of the waste products of the body, and must be removed. 3. It has absorbed water. The expired air is saturated with watery vapor. This is easily shown by breathing on a mirror, or on any polished surface. 4. It contains organic matter. The amount is usually very slight, and not easily detected. If it be in excess

it imparts a perceptible odor to the breath, which may
be both offensive and poisonous. Even with the ordinary amount, its presence is easily detected if a number
of persons be confined in a small room where there is
poor ventilation. Under such circumstances the odor
becomes very offensive, remaining in the room for hours
after it is vacated by the people. 5. The expired air is
usually warmer. To be more exact, it is generally
about the temperature of the body, being unaffected by
the variable temperature of the inspired air.

The Lungs as Excretory Organs. An excretion consists
of waste products that are useless or injurious to the
body, and must be separated, or thrown off, from it. Certain organs are known as the excretory organs, because
of the work they accomplish in this line. Viewed in this
light, the lungs must be regarded as excretory organs,
as they are constantly throwing off from the body carbonic acid and other impurities.

ALCOHOL, TOBACCO, AND THE LUNGS.

Alcohol. Persons who are in the habit of using alcoholic beverages seem to be especially liable to colds,
and to bronchial affections. Probably one explanation of
this is that often the circulation in such persons is poor,
and the heart's action affected. As a result, there is a
slight, but constant congestion of the lungs; that is, too
much blood is in the lung tissue. In some cases, late
hours and bad habits of life increase the effects just
noted.

It is certainly true that an inflamed condition of the
throat and larynx is generally found in those addicted

to strong drink, while attacks of bronchitis are not infrequent. Another fact stands very prominent. It is this: the probability of recovery from an attack of pneumonia is immensely less in the intemperate than in the temperate. We believe the great majority of physicians, everywhere, will testify to the truth of this statement. It is generally recognized that the continued use of strong drink undermines the power of the body to resist disease, and in no disease is it more marked than in the one already mentioned.

Tobacco. Tobacco especially affects the upper air passages. The irritating qualities of the smoke keep the throat congested, so that smokers suffer from a special variety of catarrh, known as the " smoker's sore throat." The larynx is often in sympathy with the throat affection, and there is a dry, hacking cough.

QUESTIONS.

1. Give three important functions of the nose.
2. State why these are so important.
3. Why is breathing through the mouth injurious?
4. What is within the larynx?
5. Where is the epiglottis? What is its use?
6. Describe the trachea.
7. Describe the air cells. What changes occur in blood here?
8. State some facts about the pleura.
9. How would you prove that the lungs have great elasticity?
10. Give the mechanism of respiration.
11. How is the chest enlarged in inspiration?
12. What is respiration? Is it voluntary?
13. How many respirations per minute?
14 How does the expired air differ from the inspired?
15. Give some of the ill effects of alcohol and tobacco.

SUGGESTIONS TO TEACHERS.

1. **The Respiratory Apparatus.** Procure the lungs of a sheep at the market. Wash in water, and wipe carefully. Call attention to the trachea, the bronchi, and the lungs.

2. **Inflation of the Lungs.** Insert a tube of any kind into the trachea. Tie the trachea tightly about it. Breathe into the tube and so force air into the lungs, thus making them expand as in inspiration. Remove the mouth and the lungs will collapse, as in expiration. A pair of bellows is the best for inflating the lungs.

3. **Lung Tissue will Float.** Cut off a small piece of the lung tissue and throw it into water; it will float, showing that air still remains in the air cells. Press the pieces in the hands to force out the air; still it will not sink.

4. **The Trachea is always Open.** Cut the trachea transversely. Try to compress it, and thus illustrate how the rings of cartilage keep it open so that it cannot collapse. Notice that the rings of cartilage are incomplete behind; and that a membrane completes them.

5. **The Bronchi.** Cut away the lung tissue which is near the lower end of the trachea. First notice how the trachea divides into two bronchi, before entering the lungs; then notice how these divide into the bronchial tubes.

6. **The Larynx.** By previous arrangement at the market, a complete larynx may be procured. This will show the epiglottis, the vocal cords, and the hard cartilaginous walls.

7. **Respiratory Acts.** Have the pupils illustrate inspiration, expiration, and a complete respiration.

8. **Respiration, Voluntary and Involuntary.** Show that respiration is voluntary by fast, then slow breathing; this soon becomes tiresome.

9. **Watery Vapor.** Breathe on any polished surface, and notice the condensed vapor.

CHAPTER XIII.

VENTILATION.

Amount of Air Inhaled. In the preceding chapter it was stated that about twenty cubic inches of air are inhaled at each inspiration. Based on this statement, and making due allowances for muscular exertion by which breathing is increased in frequency, it is safe to say each person uses at least three hundred and fifty cubic feet of air, daily, in respiration.

Oxygen supports Life. After breathing the air once it still contains considerable oxygen; after breathing this same air over and over again the amount is so reduced that the animal dies from suffocation. In the case of man, if the amount of oxygen in the air be reduced one half, breathing continues with great difficulty. A certain amount of oxygen in the air is necessary to support human life; while a much less amount of oxygen is required to support life in some of the lower animals.

It requires more oxygen for the burning of a candle than it does to support life, for a short time. Advantage is often taken of this fact to test the safety of entering a well, a vault, or an underground passage. A lighted candle is lowered into the cavity; if a sufficient amount of oxygen be present, the candle will continue to burn; if not, it will be immediately extinguished. If the candle continues to burn it will be safe for the man to enter the enclosure, for the reason that more than enough oxygen is present to support human life.

Carbonic Acid is a Poison. Each respiration not only takes oxygen from the air, but it also gives to it a quantity of carbonic acid and other deleterious ingredients. Hence it is injurious to breathe the same air, even for the second time. If the air be poor in oxygen, it will contain a large amount of carbonic acid. As this gas is heavier than the air, it will fall when confined in a small space and left undisturbed. The air which is the least capable of supporting life is then found at the lowest level. For this reason the air at the bottom of the well is much more poisonous than the air nearer the top.

An animal placed in a closed space will absorb from the air a certain amount of the oxygen, and will give off carbonic acid. Soon the surrounding atmosphere will be so saturated with carbonic acid that no more of it will pass from the body; this is according to certain laws respecting the diffusion of gases; therefore the carbonic acid is retained in the blood of the animal, causing its speedy death. This carbonic acid poisoning may occur while there is yet enough oxygen remaining in the air to support life.

From what has been said, we conclude that oxygen is necessary to life, and that an insufficient amount of it in the air will cause death. We also conclude that if the carbonic acid be not removed, it may accumulate in the air sufficiently to cause death. The air we breathe, therefore, should have a certain amount of oxygen, and should not have an excess of carbonic acid.

A person is warned when needing more oxygen and less carbonic acid, unless as the result of some accident. Headaches, restlessness, heaviness, and drowsiness result from a deficiency of oxygen and a surplus of carbonic acid. Unless the system is freely supplied with the former and readily disposes of the latter, the vital forces become lowered and the whole body predisposed to disease.

An Abundance of Air Necessary. The above facts are given especially that we may appreciate the necessity for an abundance of fresh air. It does not necessarily follow that the person must be either out of doors or in a very large room; but it does follow that, under all circumstances, an abundant supply of fresh air should be continuously within reach of the body.

Pure Air. The fact that the air is cold, and feels fresh to the face, is not proof that it is pure; currents of air may be loaded with poisons. Pure air contains the proper proportion of oxygen, and is free from poisonous gases and disease germs. Out-door air is not necessarily pure, as some sewer gas or decaying animal matter may be near. Yet, as a rule, owing to the law of the diffusion of gases, out-door air is the purest. An abundance of pure air is what is demanded.

It is not always possible to tell when the air is pure; for the best ordinary test we have is the sense of smell. Upon entering a room, if the air seems " close " it is sufficient proof that better ventilation is needed. After remaining in a close room the nose becomes accustomed to the odor and the closeness is not noticed, so the question of ventilation should be attended to as soon as the room is entered.

A Deficiency of Air Dangerous. Many cases are on record where persons have been poisoned by repeatedly breathing the same air. This is not likely to occur in our houses, because there are so many crevices about the windows and doors that enough fresh air enters. But in the holds or cabins of ships, and in the deep cells of prisons, some terrible results have occurred.

Many writers have referred to the " black hole of Calcutta " as an illustration of this fact. In a small room with only two narrow windows, there were confined one hundred and forty-six prisoners; these persons were obliged to breathe the same air over and over again, for the windows were altogether too small to allow a sufficient amount of fresh air to enter. In eight hours one hundred and twenty-three of the prisoners were dead, while those whose lives were spared endured great suffering.

Many rooms, built to accommodate large numbers at a time, have an insufficient supply of fresh air, as, for instance, schoolrooms, lecture-halls, and churches. While such a deficiency may not be enough to cause death, yet the effect on the body is marked and harmful. Drowsiness, with a dull, heavy headache, often results from a stay in such a poorly ventilated room. The listless and sleepy appearance of many a scholar is simply the result of impure air, rather than of a dull mind. " Break open the window! " shouted a noted divine, in the midst of his discourse, as he saw many in his congregation asleep. He knew that the most brilliant speaker could not overcome the drowsy effects of impure air.

If it becomes necessary to spend a considerable time each day in a poorly ventilated or overcrowded room the whole body is soon affected. Living in poorly ventilated rooms enfeebles the whole body; the appetite fails, the red corpuscles are reduced in number, the skin becomes colorless, and the whole system shows that it is suffering from too little oxygen and from too much carbonic acid. Colds and coughs are frequent; and the system has only slight power to resist disease of any kind. The whole condition is one of " oxygen starvation " and carbonic acid poisoning.

The Proper Amount of Air. It has been stated that an abundance of pure air is necessary. It is better to say that an over-abundance is necessary, so that if any error be committed it may be on the safe side. The supply of fresh air required for a room depends largely on the number of persons in the room; for it is evident that a hundred persons will require a hundred times as much air as one person.

Then, too, the presence of fires in the room must be considered. The burning gas consumes much oxygen and gives off carbonic acid. For this reason a crowded hall, on a cold winter's evening, with heavy fires and lighted gas, requires much more ventilation than when a small company is assembled on a warm summer's day. As it would cause discomfort to raise the windows and open the doors during the winter time, it follows that some system of ventilation is absolutely necessary for all places where persons are likely to assemble.

How to Obtain Pure Air. How to obtain an unfailing supply of fresh, out-door air in our rooms is the constant study of those who plan homes and public buildings. Years ago this subject received no attention whatever. This was partly due to the fact that the method of heating the houses was far different from that used at the present day. The open fireplaces made a constant change of air, while the cracks about the doors and windows furnished avenues through which the fresh air entered the rooms.

A furnace gives good ventilation, because as the warm air enters the room from the registers the cold air escapes through ventilators in the wall near the floor; thus a constant current of air is established. Great care should be exercised to see that the cold air-flue of the furnace receives its supply from clean surroundings. If the out-door air, entering the flue is near decaying animal or vegetable matter, or an imperfect sewer, then the impure air might be conveyed into the building, producing severe and perhaps fatal sickness.

If the rooms be heated by a furnace, the air should be moistened by having it pass over a dish of water. A failure to do this results in the necessity of breathing very dry air, which is decidedly injurious. The open grates of modern days are good ventilators. An ordinary stove is a means of ventilation, for as the draught passes through it and up the chimney, fresh air comes in through the opening of doors and the crevices of the windows to take its place.

Attention should always be given to the ventilation of the sleeping rooms. One third of our entire lives is spent in these rooms, yet how often do we neglect to make them either cheerful or healthful. Unless the builders of the house have provided some method of ventilation, the air may be changed by raising the lower sash of one window and lowering the upper sash of another. A better method, however, is to admit the air into the room through wire gauze, used as window screens. There are a number of ventilators for sale in the market which allow a free passage of air and yet effectually prevent draughts.

If the lower window sash be raised about six inches and a board be placed under it, completely filling the space between it and the window casing, there will be established some ventilation between the sashes where they meet but where they no longer fit tightly. This is a fairly good method for the winter, but hardly sufficient for the more quiet air of summer.

The teacher generally gives personal attention to the ventilation of the schoolroom, and the proper authorities should insist that some method be devised in order that the change of air may be constant and abundant. This does not mean ventilation at noon and recess only; it means that the change should be continuous and uninterrupted, for anything short of this fails to answer the purpose.

Improper Ventilation. Ventilation is improper if it causes sudden changes in the temperature of the air. After the hearty plays of recess when the skin is moist with perspiration, the pupils should not sit down in a cool room. This is too often the cause of colds and coughs. The schoolroom should be of an even temperature all through the day; therefore there must be a constant and uninterrupted change in the air. Ventilation is improper if it produces a current of air; if a draught of air be allowed to strike the back of the neck, or any sensitive part of the body, it is very likely to cause a cold. These currents of air should be carefully avoided, especially when the body is resting from active exercise.

Night Air. Some persons seriously object to opening the windows of their sleeping rooms at night, for fear of "the deadly night air." Yet all the lower animals breathe it, from the delicate and tender young to the strong and aged. Soldiers and hunters breathe it as they sleep beneath their tents, and in the open air, while many invalids have been restored to health by living out of doors both day and night. Fear cold draughts, but do not fear the night air. Florence Nightingale said : —

"The choice is between pure air without and impure air within; most people prefer the latter, an unaccountable choice. An open window most nights in the year can hurt no one. In great cities the night air is the best and purest in the twenty-four hours. I could better understand, in towns, shutting the windows during the day than during the night."

Ventilate the Cellar. The cellars of houses and other buildings are often great reservoirs of foul air. The cellars of dwellings frequently have stored in them quantities of vegetable matter which give off injurious gases as they decompose. As the room is usually dark, the decaying organic matter is unseen, and hence it remains until the escaping gases penetrate the rooms above, and endanger the lives of their inhabitants. Cellars should be well ventilated, kept scrupulously clean, and so built, if possible, that the sunlight can enter them.

Sewer Gas. The escape of gas from defective plumbing of houses, and from improperly constructed drains and sewers, is the cause of much sickness. Gas may escape from a sewer which is near or under the house, and enter the rooms without being detected. The first intimation of its presence may be a severe case of diphtheria, or some low form of fever. No one should occupy a house, no matter how well it may be provided " with all modern improvements," until he is satisfied that the plumbing is in good condition, and that the sewers and drains are properly constructed.

Deodorizers. One odor may cover another without destroying it. A free use of cologne may cover the odor of a poorly ventilated room, but it will neither remove the carbonic acid and organic matter, nor will it bring more oxygen. Coffee and sugar are often burned in a room to destroy some poisonous or disagreeable odor. They do not destroy, however; they simply cover one odor with another which is more powerful. Any substance that will replace or cover the odor of another, and yet not destroy it, is called a deodorizer.

Disinfectants. There are substances which actually destroy odors; these are true disinfectants. Bad odors in the atmosphere often depend upon the presence of impurities dangerous to health and life; these must be removed, as far as possible, by free ventilation and by the use of destroying agents. Many chemicals possess the power of purifying the air from germs, and from the products of decaying animal and vegetable matter. Disinfectants are used to purify sewers, cess pools, sinks, and to destroy the germs of scarlet fever, diphtheria, and small-pox. They are also largely used by Boards of Health and by physicians, to destroy disease germs. A disinfectant may be perfectly odorless itself, and yet have the power of destroying the most offensive odors. The chlorides and sulphates of the metallic salts are powerful disinfectants. Preparations of the chlorides are on the market which are reliable and convenient.

The sulphate of iron (copperas), dissolved in water in the proportion of four ounces to the gallon, is a useful disinfectant for cleansing gutters, drains, sewers, etc.

Absorbents. Absorbents are used to take up the gases from decomposing materials. Lime and charcoal are the most frequently used for this purpose. White-washing a room renders the air sweeter and purer because it absorbs certain gases in the atmosphere.

Contagion. Some of the most poisonous substances are entirely free from odor. This is the case with the germs which are believed to cause a number of diseases. Scientists are not agreed as to the germ theory of disease, but it seems altogether probable that a number of diseases, such as diphtheria, scarlet fever, measles, and small-pox are caused by minute germs. In some way the germs gain an entrance into the body where they develop and cause disease.

If a well person approaches one thus ill and comes in contact with these germs, and if the body of the well person be in just such a condition that these germs can there thrive and develop, then the exposed party will, in due time, be ill with the same disease. These germs may lie dormant for years until their surroundings are suitable for their development. A grain of wheat may be kept for years and show no signs of life; but when at last it is surrounded with warmth and moisture it begins to revive and to give evidences of life. Soon a sprout appears, when, if food and sunlight be added, it will bear fruit.

Thus disease germs may remain in a room for a long time, clinging to the paper, the carpet, or the walls. Months, or even years afterwards, some inhabitant of the room may fall a victim to scarlet fever or small-pox. Then it is remembered that not since the illness of the last patient has the room been occupied, and never has it been thoroughly disinfected. Germs may not only lie dormant in this way, but they may be carried long distances in the clothing, or in the hair.

From the above outline, only too brief for such an important subject, some valuable conclusions may be drawn. Persons suffering from such contagious diseases as scarlet fever or diphtheria should be kept in a well ventilated room, which none but the physician and nurse should enter; the nurse should leave the room only for exercise, and then she should have a second room in which she could change her clothing, and make free use of the best disinfectants. These ideas are now thoroughly understood by all competent physicians, who can be relied upon to see them carried out in detail.

QUESTIONS.

1. Give the amount of air inhaled with each inspiration.
2. What is said about the necessity of oxygen?
3. What do you learn about carbonic acid?
4. Pure air should contain what? And be free from what?
5. Give some of the bad effects from a deficiency of air.
6. The supply of fresh air required for a room depends upon what conditions or circumstances?
7. Give some of the methods mentioned for ventilation.
8. When is ventilation improper?
9. What is a deodorizer?
10. What is a disinfectant? Name one.
11. Name some absorbents.

CHAPTER XIV.

THE KIDNEYS.

General Description. The kidneys are two in number, one on either side of the spinal column. Each kidney is about four inches in length, two inches in breadth, and about an inch in thickness, and from four and one half to six ounces in weight. The upper border of the kidney is about on a level with the eleventh rib.

A kidney resembles a bean in shape. It is completely covered with a thin membrane, called the capsule. The convex border of either kidney is placed toward the side of the body, while the concave border is next the spinal column. Each kidney is supplied with blood by an artery which arises from the aorta. This vessel enters the kidney as shown

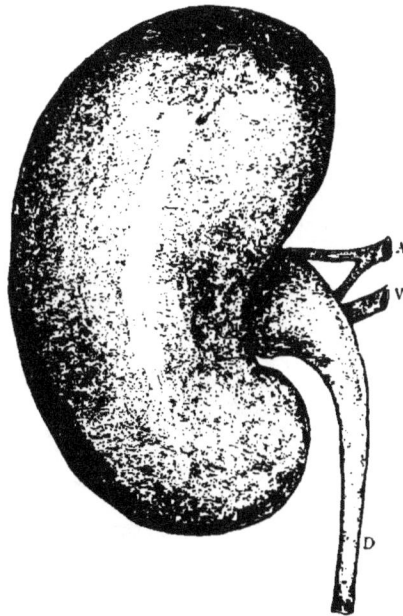

FIG. 45. A kidney : A, an artery ; v, a vein ; D, the duct that carries away the materials filtered from the blood.

at A, Fig. 45. After the arterial blood has circulated through the kidney it is returned through a vein at V. This vein empties into the large vein which lies by the side of the aorta.

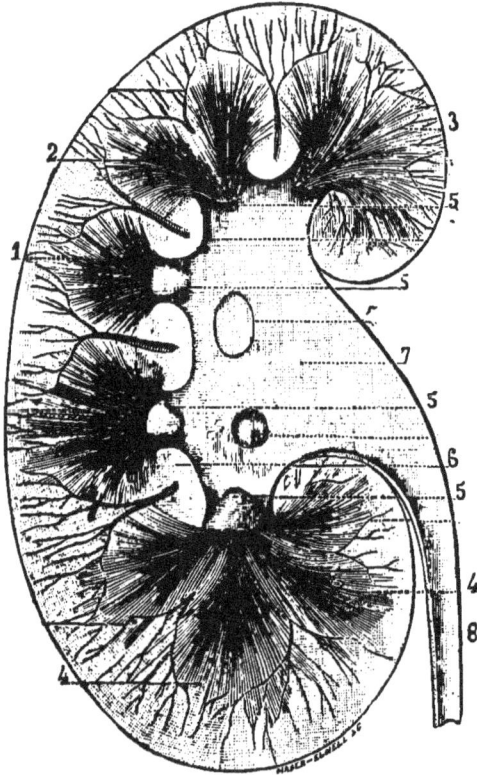

Minute Structure. On making a section through the long diameter of a kidney, as represented in Fig. 46, it appears to consist of two different substances. The outer part, near the convex border, looks red and granular. Farther in, are seen a number of little eminences, illustrated at 5, Fig. 46. Fine lines extend fan-shaped, toward the convex border. They represent a collection of canals.

FIG. 46. A section through a kidney: (1, 2, 3, 4), collections of tubes, or canals; (5), papillæ on which the tubes open; (6), below the end of the dotted line is a blood vessel; others are seen above this; (7), the dilated beginning of the duct, (8), which carries the secretion from the kidney.

The most interesting portion of the kidney is found in that part near the convex border, to the left of Fig. 46. Here active changes are constantly taking place. In this portion there are seen, with the microscope, vast numbers of small, round, red bodies, which are but the beginnings of long, narrow tubes. Two of these red bodies are illustrated in Fig. 47. Each body consists of coils of capillary blood vessels, surrounded by a membrane. While the blood is circulating through these bodies, certain substances, principally water, are taken from the blood. These substances are carried away through the minute canals until they empty into a large duct, shown at 7 and 8, Fig. 46. This duct carries the secretion directly to a reservoir, designed especially for its reception.

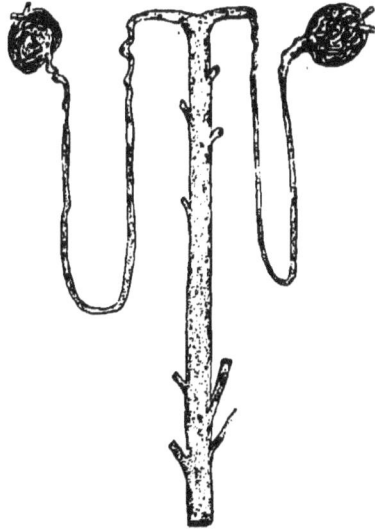

FIG. 47. One of the tubes of the kidney, magnified.

Thus we find that the kidneys are purely excretory organs, taking from the blood materials which would rapidly prove poisonous if allowed to remain in the body.

Urea. The most important ingredient taken from the blood by the kidneys is called urea. It represents the worn-out, or used up, nitrogenous tissues of the body. A greater part of the nitrogen which enters the body with the foods is removed from it by the kidneys in the form of urea. If, for any reason, the kidneys fail to take the urea from the blood, the most serious results follow. Symptoms of blood poisoning soon appear, and convulsions and even death speedily follow, unless the difficulty be promptly relieved.

If some of the tubes be affected, as in Bright's disease, then the kidneys do not completely remove the waste products from the blood, and prolonged sickness results. At last, the accumulation of urea in the system is sufficient to produce poisoning and death.

Danger from Colds. Disease of the kidneys is often caused by exposure to cold, especially if the body be moist with perspiration at the time. If the function of the skin be suddenly checked by the action of cold, the blood leaves the surface and goes to internal organs, as the kidneys, causing congestion ; the excretory work the skin was performing is thus suddenly thrown on the kidneys, making them do double duty. As a result they become inflamed and unable to properly perform their work.

ALCOHOL, AND THE KIDNEYS.

Alcohol is recognized as one of the most frequent causes of kidney disease, and especially of that form known as Bright's disease. When much alcohol is used, a certain proportion of it passes out of the body unchanged, through the kidneys. The alcohol comes in contact with the delicate cells lining the tubes, and so alters them that they are unfitted to do their work.

If the use of the alcohol be continued the cells gradually diminish in size, the round bodies become smaller, and the tubes and canals likewise are reduced. The connective tissue, which is used only to hold the tubes and canals together, grows thicker and thicker, until the whole organ becomes largely composed of it.

At last many of the tubes become so useless that the urea is not taken from the blood, and death occurs. To produce these changes in the structure of the kidney it is not necessary that the drink be strong with alcohol. Many such cases are directly caused by the use of ale and beer.

It is absolutely necessary that all poisonous materials be removed from the body as nature intended. We may deprive ourselves of food and keep nourishment from entering the body for days at a time. We may, in this way, check the action of those organs directly concerned in the secretion of the digestive juices, as the stomach and the pancreas. But we cannot check, even for the shortest time, the action of those organs employed in the removal of the worn-out and poisonous materials of the body, without seriously endangering the health and even the life.

In view of these facts, it does not seem possible that any one who is familiar with them would try the experiment of taking alcoholic beverages into the body ; especially so when science so clearly points to their direct effect on such important organs as the kidneys. It is an experiment that has cost many thousands of lives.

QUESTIONS.

1. Give a general description of the kidneys.
2. What is the most important ingredient taken from the blood by the kidneys?
3. What does the urea represent?
4. Explain how exposure to cold affects the kidneys.
5. Give some of the effects of alcohol on the kidneys.

CHAPTER XV.

THE BONES.

General Description. There are two hundred and four distinct bones in the body. This does not include the teeth, the patella or knee-pan, and a few other bony structures occasionally present. The bones vary greatly in form; some are long and stout, as the femur; others are short and small, as the bones of the hand and foot. Some are flat, as many of the bones of the skull, while others are so irregular that it would be very difficult to describe them. On closely examining one of the larger bones, eminences and depressions are seen. The eminences afford places for the attachment of muscles and tendons, while the depressions afford passages for blood vessels, nerves, and tendons.

FIG. 48. A bone with the periosteum partly peeled off.

The Periosteum. A thin membrane called the periosteum, surrounds each bone. It is composed of two layers, an outer layer of firm tissue, which is simply for support and protection, and an inner layer of cells. This inner layer is essential to the life of the bone, and its cells are even capable of forming new bone. In cases where it is necessary to remove a portion of a bone, the surgeon avails himself of this fact. He simply peels back the membrane, removes the injured bone, and thus leaves the periosteum to make new bone.

The periosteum is well supplied with blood vessels, some of which pass directly into the bone through minute openings on the surface. Many of these openings can be seen on any bone from which the periosteum has been removed.

Compact and Cancellous Tissue. If one of the long bones of any animal be sawed lengthwise, it will be found to be hollow, except at the ends. The hard, thick wall, midway between the ends of the bone, is called the compact tissue; while the spongy, honey-combed bone,

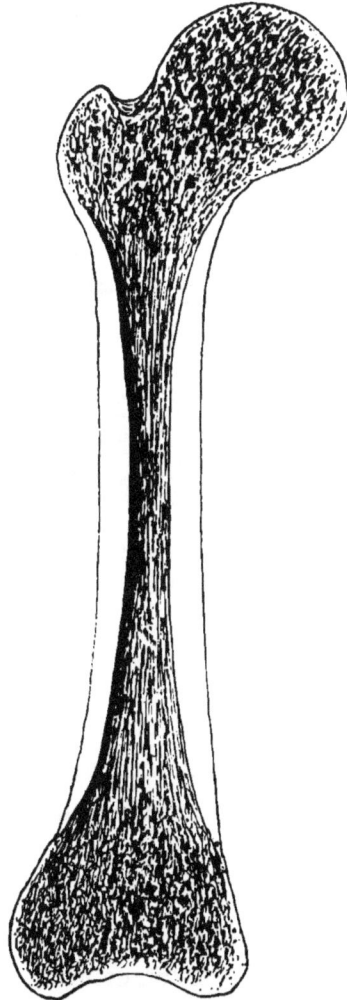

FIG. 49. Longitudinal section of the femur. The spongy, or cancellous bone shows at the ends; while the compact, hollow bone forms the shaft.

occupying the center of the ends of the bone is called

the cancellous tissue. The smaller bones and the flat
bones are not hollow ; they have an outer layer of compact bone, within which is the spongy bone, or cancel-
lated tissue.

Fig. 50. Transverse section of the femur. The section to the left is from
the shaft ; to the right is from the head or upper portion of the femur.

The Marrow. The large central cavity of the long
bones, and all the spaces in the spongy bone, are filled
with a yellowish or reddish substance, called marrow.
It consists principally of fat cells and marrow cells.
These marrow cells doubtless give rise to some of the
red corpuscles of the blood. There are other sources of
origin for these corpuscles, but it is well established that
some of them originate in this way.

Microscopic Structure. If a longitudinal piece of bone
be ground very thin, and then examined with a microscope, a number of canals will be seen parallel to the
long axis of the bone. These canals unite with each
other, as shown in Fig. 51, by short branches extending
across nearly at right angles. A cross section of bone
shows that these canals are circular or oval in shape, as

illustrated in Fig. 52. These are the Haversian canals, so called after their discoverer, Cloptin Havers. In the living bone all these canals are filled with blood vessels, while the frequent communication of the canals permits the blood to flow to all parts of the bone. But the most

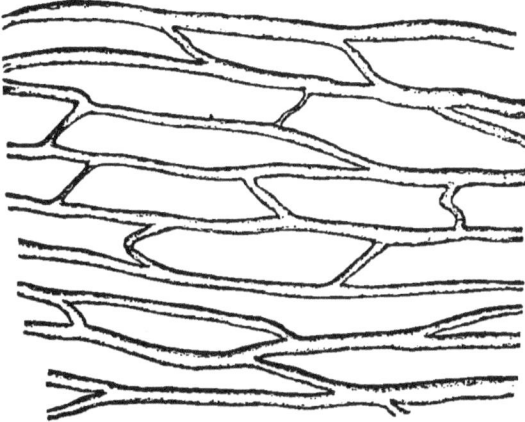

FIG. 51. Longitudinal section of bone, showing the Haversian canals, magnified.

interesting fact connected with the structure of bone is that living cells are always present in them.

The Bone Cells. Between the Haversian canals there are oblong spaces, called lacunæ, signifying a hole or a lake. Extending from the lacunæ in every direction, are minute canals, called canaliculi. In these irregular-shaped spaces are the bone cells. In Fig. 53 is a longitudinal section of bone showing these holes, with their many-branched canals extending from them. Fig. 52 is not so highly magnified, but it represents the vast number of these lacunæ, with their canals leading from them.

The lacunæ are very small. yet the canaliculi are much smaller, being not over $\frac{1}{20000}$ of an inch in diameter. Yet in all these there is living matter,

the bone cells.   These cells are nucleated, and are of
the same shape as are the spaces in which they rest.
Therefore the idea that bone is dead tissue, not subject
to change, must be abandoned.   Bone is a living tissue,
filled with blood vessels, nerves, and cells, which are as
much endowed with life as are the cells of other parts
of the body.   A framework of fibers extends throughout
the whole bone, holding all the parts in place.

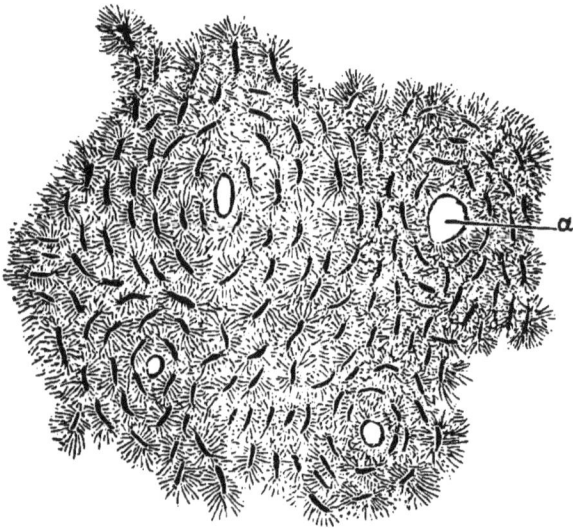

FIG. 52.   Transverse section of bone, not as highly magnified as
Fig. 53.   Four Haversian canals (*a*) show, surrounded by lacunæ and
canaliculi.

Animal and Mineral Matter.   The parts just described
— the blood vessels, bone cells, marrow, nerves, and the
framework of fibers — constitute the animal matter of
bone.   But the body must have a strong support, some
kind of a framework to which muscles can be attached,
and in which the most delicate organs can be protected.
We find, therefore, that some mineral substance is added

to the soft and yielding animal matter; this makes the whole tissue hard and unyielding.

By remembering that fire will destroy animal matter and that acids will dissolve mineral matter, some interesting results may be obtained. For instance, if a fresh

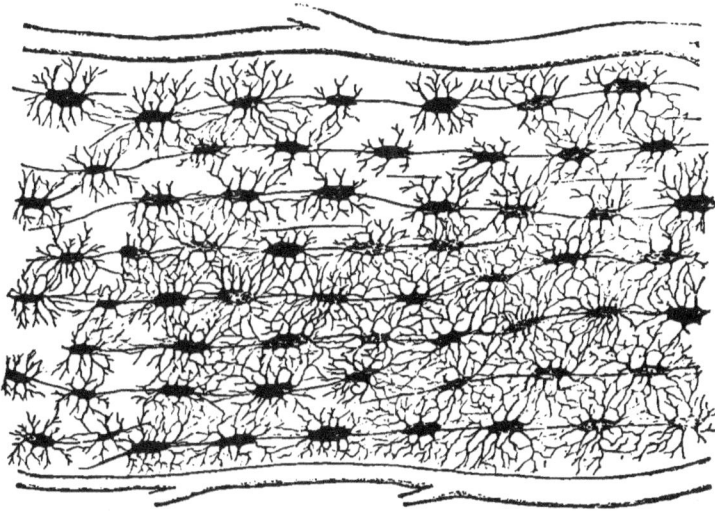

FIG. 53. Longitudinal section of bone, highly magnified. Between the two Haversian canals are a number of lacunæ and canaliculi.

bone be placed in the fire, and subjected to heat for a considerable time, all the animal matter can be burned out. The shape of the bone will not be changed; it will only become lighter and whiter. After such treatment it can be easily broken and pounded into a fine powder.

The mineral matter consists largely of lime. This is easily removed by soaking the bone in a weak acid for a few hours. The shape of the bone will not be changed; it will only lose its hardness, and become easily bent in any direction. All the blood vessels and the bone cells still remain in the bone.

11

From these experiments we easily draw the conclusion that, if the bones do not contain the proper amount of mineral matter, they will bend, and will be unable to keep their shape, and properly support the tissues around them. If there is a deficiency in the amount of animal matter, the bones will be too brittle, and liable to break if any extra strain be brought upon them.

Nature very wisely provides that in early life there should be an excess of the animal matter in bone. If this were not so the tumbles and falls which are the common lot of all children at play, would result in many serious injuries. But the bones of youth have such a spring to them that children are not likely to suffer often from their fracture. In old age the opposite condition exists, and there is an excess of mineral matter. The bones are very brittle, and are much more easily broken; a slight fall often being sufficient to fracture a large bone.

The Reproduction of Bone. It has already been stated that bone is a living tissue. Because of this, it is capable of being reproduced if only some of the original cells be left undisturbed. If the end of a young bone be removed it may again be reproduced so that a fairly good joint will be formed. Pieces of bone that have been broken, or sawed off will again unite with the original bone.

We have already mentioned that if a portion of bone be removed and the periosteum left, new bone will soon take its place. The periosteum, even when transplanted to another part of the body, will give rise to new bone. If a large bone be fractured, the periosteum soon deposits a substance on the broken ends. This is at first soft, and of jelly-like consistence; later it becomes more solid and resembles cartilage ; and still later thin layers of bone appear, reuniting the broken ends. As hard and firm as bone appears, physiology teaches that it is subject to the same laws of repair and waste, growth and decay, as are the other tissues of the body.

A Good Form. To have a fine, erect figure is certainly desirable. But this cannot be secured if certain laws of nature be persistently broken. The body must not be distorted by improper dress, or by wrong methods of walking or sitting. The figure is easily made crooked by repeatedly yielding to a feeling of languor, or by sitting in a cramped and unnatural position. In standing it is better to rest the weight on both feet. The habit of resting the weight on one and the same foot is sure to make the hip bones grow out of shape ; it will also bend the spine, and make it incline toward one side. In walking, the whole body should be erect, with the shoulders well thrown back.

QUESTIONS.

1. Give a general description of the bones.
2. What is the periosteum ?
3. Where is compact bone found ? Where cancellous ?
4. Where is the marrow found ? Of what is it composed ?
5. Describe the bone cells.
6. What constitutes the animal matter of bone ?
7. How does fire affect bone ?
8. How can the mineral matter be removed ?
9. What conclusion can be drawn from these experiments ?
10. During what period of life do the bones contain an excess of animal matter ? When an excess of mineral matter ?
11. How is bone reproduced ?

CHAPTER XVI.

THE SKELETON.

Object of the Skeleton. All the higher animals are provided with a skeleton. This is used either as a support, or framework, for the organs and tissues of the body, or as a protection from injury. In some of the lower animals the skeleton is entirely on the outside. The oyster is completely inclosed in its hard shell, and is thus well protected against the attacks of enemies. The lobster has an exterior skeleton also, but the parts are so arranged that there is considerable freedom of motion. The turtle not only has an interior skeleton, but also a large plate, or exterior skeleton of hard material. This animal can withdraw his head beneath the outer skeleton, and thus the whole body is protected from violence.

The animals which have exterior skeletons do not have such freedom of motion as is required by the higher animals. Therefore, in man, beasts, birds, fishes, and some other animals, the skeleton is entirely within the body. It gives to the body a solid framework, to which can be attached ligaments for holding the joints together, and muscles for moving the various parts. It also makes a more or less complete covering, or protection, to many important organs. As a protection, the bones of the skull furnish a perfect covering to the delicate texture of the brain ; they nearly surround the eye ; while the most delicate parts of the ear are deeply imbedded in bony tissue. The ribs, spinal column, and sternum make a nearly complete covering for the heart and lungs.

Bones of the Skull. The skull is composed of twenty-two bones, which are usually divided into the bones of the cranium, and the bones of the face. The bones of the cranium are eight in number, and they are so arranged that they form a solid and strong covering for the brain. At the base of the brain is a large opening through which passes the spinal cord and large blood vessels. The bones of the face are fourteen in number ; they protect the organs of the special senses of sight, smell, and taste, and they also provide for the reception of the teeth.

The skull rests upon the first vertebra and upon a tooth-like projection, or pivot, of the second. It nods, or moves forward and backward, upon the first vertebra, which remains stationary. It moves from side to side by the motion of the first vertebra upon the second. During this motion the skull and the first vertebra move together, the vertebra swinging around the pivot, which extends upward from the vertebra below it. A glance at Figs. 53, 54, and 55 will aid in making this clear.

The Spinal Column. The spinal column consists of twenty-four small bones, — resembling those illustrated at Figs. 56 and 57, — and two irregular bones, at the lower end of the column. The two irregular bones, at an early period of life, are composed of nine separate pieces, each representing a vertebra. Later, five of these unite to form one bone, and four to form another. These two compound bones form the back of the pelvic cavity.

A reference to Fig. 54 shows that the spinal column is not straight, but forms a series of curves. In the neck, and in the abdomen, the convexity of the curves is forward, while in the chest, and in the pelvis, it is backward. These curves give additional elasticity to the column, and with the aid of the elastic cartilages give great protection to the brain from sudden jars.

Each bone in the spinal column is called a vertebra, from a Latin word signifying to turn, as a joint. Thus a translation or definition of the word would be, "a joint of the spinal column." Each vertebra has within it a large opening, through which passes the spinal cord. The vertebræ are held together by ligaments, and are so placed, one directly over the other, that the central openings form a continuous canal extending the entire length of the spinal column. This is called the spinal canal; it furnishes a

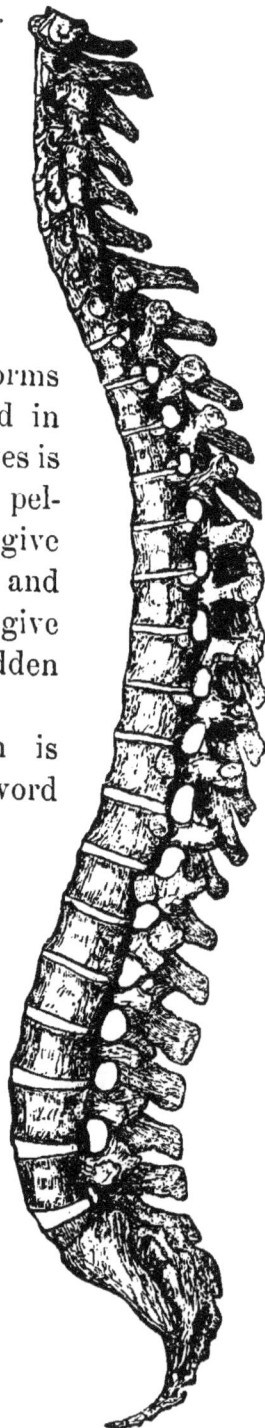

FIG. 54. The spinal column. The right side of the figure is toward the back of the body.

firm protection to the spinal cord. Between the vertebræ are discs, or cushions, of elastic cartilage. This cartilage resembles rubber in its elasticity. Its great use can be appreciated when it is stated that the combined thickness of all these cushions is over six inches. They greatly diminish the shock and jar that comes to the body from jumping and running.

Nearly all the vertebræ resemble those shown in Figs. 56 and 57. Fig. 56 illustrates a vertebra of the neck, as viewed from above. The long process, C, is the one that is so

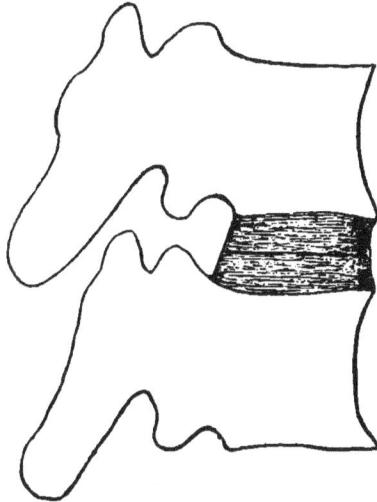

FIG. 55. Two vertebræ, with the elastic cushion of cartilage between them.

easily felt at the back of the neck. The darkly shaded oval portion at the top of the figure, immediately in front of the opening for the spinal cord, is the place of attachment of the elastic cartilage. Fig. 57 represents a vertebra lower down the spinal column, and viewed from the side. A represents the front of the vertebra, and the place where the cartilage is attached. The long process, B, extends backward and downward, forming a part of the ridge which may be felt extending down the center of the back.

The "atlas" is so named because it supports the globe of the head. It stands at the top of the spinal column, and differs in shape from the other vertebræ.

168   THE ESSENTIALS OF HEALTH.

It has no cushion of cartilage to separate it from the bones above or below it, neither has it any long process.

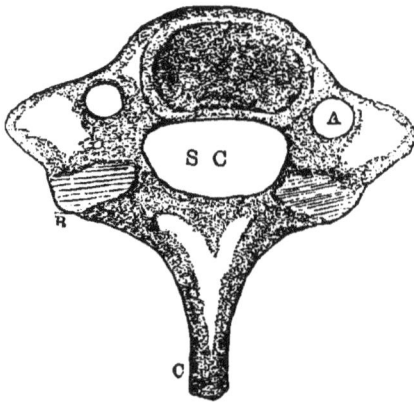

There are two places on the atlas, however, which are of interest; they are illustrated at 3, Fig. 58. On these surfaces the base of the skull rests and moves. It has a large opening in its center, which is divided into two parts by a strong ligament, illustrated in Fig. 58 by a dotted line. Through the back and larger opening, 2, the spinal cord passes; while through the smaller

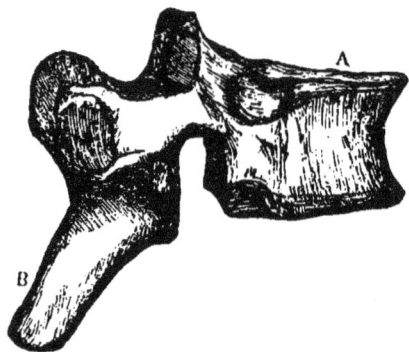

FIG. 56. The upper surface of one of the vertebræ of the neck: A is an opening in each side for a blood vessel; B is the point on which the bone above it rests; c is the long process that forms a part of the ridge which extends from the back of the spinal column.

opening, 1, passes a bony pivot, projecting up from the bone directly beneath.

The "axis" is so named because it forms the pivot upon which the head turns from side to side. Its most peculiar feature is the strong bony projection, resembling a tooth, which rises perpendicularly from its upper and front part. When the atlas and the axis are in position

FIG. 57. One of the vertebræ viewed from the side: A, represents the body of the vertebra; B, the process.

they correspond with Fig. 60. A study of these last three illustrations will make it clear how the head turns from side to side with the atlas, while the axis is stationary,—both the head and the atlas rotating on the pivot of the axis. They also show how the head can bend forward and backward, while the atlas remains fixed and immovable upon the axis. It is one of the most curious and wonderful mechanisms of the body.

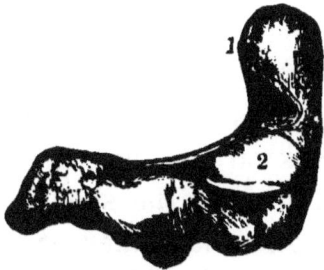

Fig. 58. The atlas, or the first vertebra, viewed from above : (1) the process of the axis; (2) the opening for the spinal cord; (3) the places on which the skull rests. The dotted line represents a ligament which holds the process, (1), in place.

The Ribs. There are twelve ribs on each side of the body. The ribs are so curved that each makes an elastic arch of bone. Behind, the ribs are attached to the spinal column. In front, the first seven are attached to the sternum, by means of cartilages. The next three are fastened to each other by cartilages; while the last two have no attachment in front, hence they are called the floating ribs.

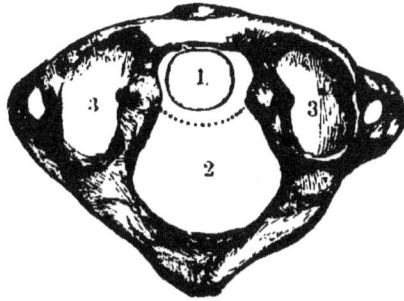

Fig. 59. The axis, or the second vertebra, viewed from the side : (1) the process on which the atlas turns seen also at 1, (Figs. 58, and 60) ; (2) the place on which the atlas rests.

The Thoracic Cavity. The thoracic cavity is inclosed by the spinal column behind, the sternum in front, the ribs on the sides, and the diaphragm below. The spaces between the ribs are filled in with muscular tissue, so that surrounding the thoracic cavity there is a complete wall, formed partly of bone and partly of muscle. All that part

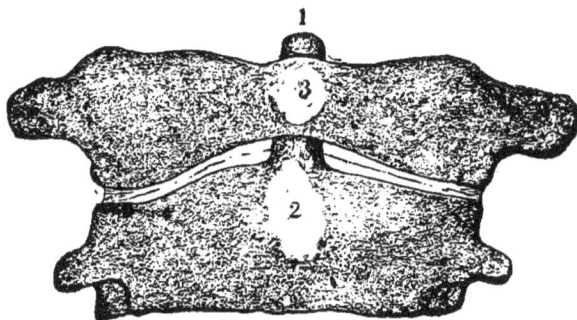

FIG. 60. The atlas and axis in position, front view : (1) the tooth-like process of the axis, showing above the atlas; (2) the axis; (3) the atlas.

of Fig. 43 which is above the diaphragm, 6, represents the thorax, or chest, with its contents in place. In this cavity are the lungs, heart, and large blood vessels.

The Upper Extremities. There are five large bones and several small ones that belong to the upper extremity. The clavicle, or collar bone, extends from the front of the shoulder to the top of the sternum ; it keeps the shoulder joint outward and backward. The scapula, or shoulder blade, forms the back part of the shoulder, and is between the shoulder joint and the spine ; at its upper and outer part is a cavity, or socket, for the reception of the head of the humerus. The humerus, or arm bone, extends from the shoulder to the elbow. Its head is round, forming a ball for the ball-and-socket joint at the shoulder; the lower end is flattened and grooved, to make the hinge joint of the elbow. The radius and ulna are the two bones of the forearm.

The ulna forms the hinge joint at the elbow with the humerus, while the radius forms the upper part of the wrist joint. The radius and ulna are placed side by side, and are so arranged that the radius can move

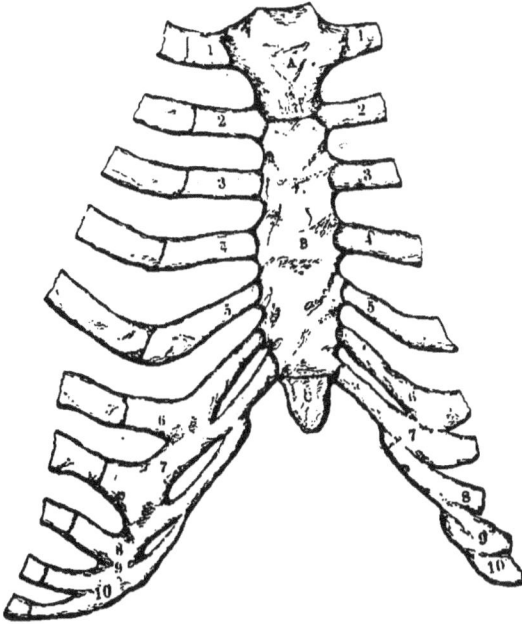

FIG. 61. Front view of the sternum and parts attached to it; (A, B, C) the three portions of the sternum; (1, 2, 3, 4, 5, 6, 7) the cartilages which unite the first seven ribs directly to the sternum; (8, 9, 10) the cartilages which unite the eighth, ninth, and tenth ribs indirectly to the sternum. To the right of the Fig. only the cartilages are shown; to the left, the ends of the ribs are shown also.

partly around the ulna, giving the hand the power of rotation, or turning. The radius is on the side of the arm corresponding with the thumb. The remaining bones of the upper extremity form the hand, and by their arrangement allow a great variety of movements.

The Pelvis. The pelvis is inclosed, on the back, by the two lower bones of the spinal column; on the sides, by the large hip bones; and in front, by the pubic bones. The curved shape of the upper edge of the hip bones gives a strong support for the abdominal organs. The sides furnish a means of attachment for the lower extremities, while the back gives attachment to the spinal column.

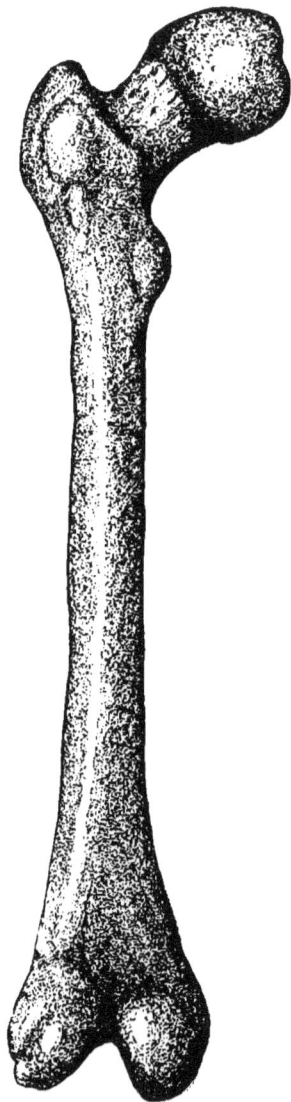

The Lower Extremities. The bones of the lower extremity are much like those of the upper. The femur, or thigh bone, is the longest, largest, and strongest bone in the body. Its head is ball shaped to form the ball-and-socket joint with the bones of the pelvis. Near the head is a roughened mass of bone, to which muscles are attached. The lower end of the bone is flattened, to form, with the tibia, the hinge joint of the knee. The tibia and fibula are the two bones which form the leg. They are placed side by side, but are not movable as are the bones of the forearm. The front of the tibia forms the sharp ridge felt in the front of the leg, while the fibula forms the out-

FIG. 62.. The femur.

side of the leg. The patella, or knee pan, protects the knee in front.

Bones of the Foot. The bones of the foot are so united that they form an arch, of which only the front and the back touch the ground. This arch is useful in protecting the body from severe shocks, as in the act of running or jumping; for it is evident that when the weight of the body is thrown upon the arch,

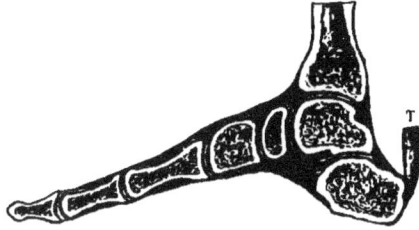

FIG. 63. Side view of the bones of the foot, naturally arranged in the form of an arch.

its center is pressed downward, thus acting as a spring.

The Joints. Whenever bones form a joint they are covered with a layer of highly polished cartilage. This gives some slight elasticity, and also reduces the friction. Covering the cartilage is a very thin membrane which is constantly secreting, or pouring out, a watery fluid, called the synovial fluid, or joint water. It serves the same purpose as does oil to the wheels and joints of machinery.

The Ligaments. The bones are held together at the joints by bands of tissue, called ligaments. These are very dense and strong, and capable of withstanding

Fig. 64. A longitudinal section through a joint, showing a layer of cartilage over the end of each bone: (1) the ends of the bones; (2) the layers of cartilage.

great strain without injury. Fig. 65 shows how firmly and completely the shoulder joint is covered with these ligaments. Fig. 66 illustrates how the head of the femur is buried in the socket prepared for it; while Fig. 67 illustrates the hip joint after the outer ligaments have been removed, and the bone pulled partly out of its socket. A strong ligament still remains, holding the head of the femur to the center of the socket.

Sometimes the ligaments are unduly stretched, or slightly torn, as when the wrist or ankle is sprained. Such an injury usually causes great pain, and recovery is slow. The ligaments may be so broken or torn that the bones slip out of their proper places. The bone is then said to be " out of joint," or dislocated. In a few healthy persons the ligaments are very loose, so that, by the action of the muscles alone, some of the joints can be dislocated at will. Such persons are said to have loose joints.

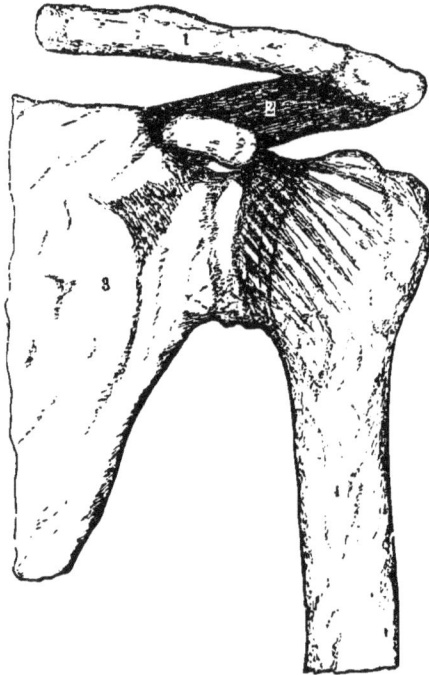

FIG. 65. Front view of the bones of the shoulder: (1) the clavicle; (2) a process which extends from the back of the scapula, and which can be felt as a prominence on the back of the shoulder; (3) the scapula; (4) the humerus. The joint is enclosed by ligaments.

Varieties of Joints. Joints may be either movable, im-

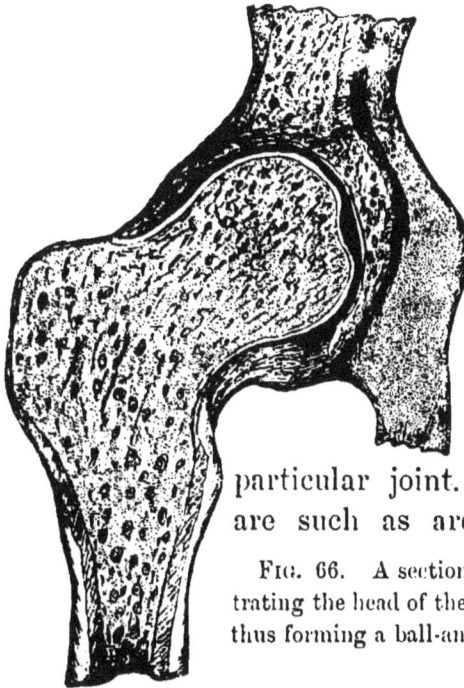

perfect, or immovable. The movable joints vary exceedingly in the degree of their mobility. The opposing bony surfaces move upon one another, and the extent of the motion is only limited by the structure of that particular joint. The imperfect joints are such as are found in the spinal

FIG. 66. A section through the hip joint, illustrating the head of the femur resting in a socket, and thus forming a ball-and-socket joint.

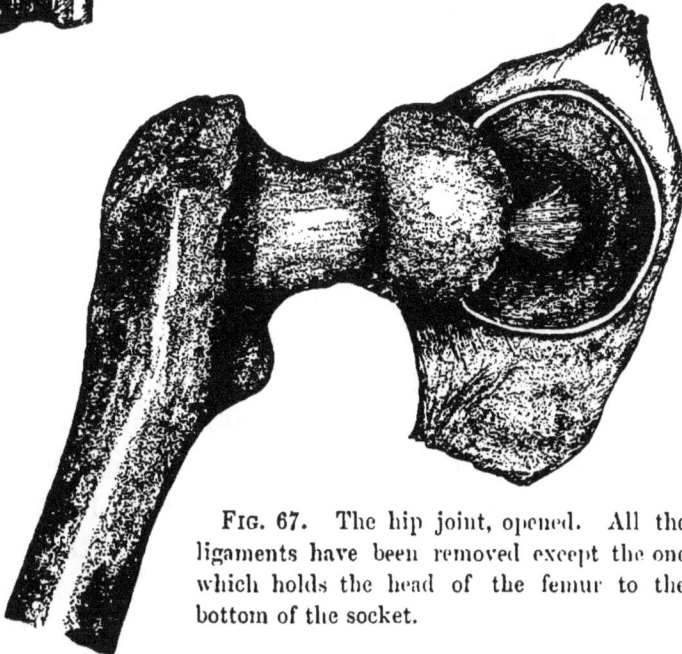

FIG. 67. The hip joint, opened. All the ligaments have been removed except the one which holds the head of the femur to the bottom of the socket.

column. There is some slight motion between the vertebræ, due to the elasticity of the thick plates of cartilage. It is the same motion that can be made after cementing a thick plate of rubber between two blocks of wood. The elasticity of the rubber would allow a certain freedom of motion and yet the parts would not move upon one another. Examples of the immovable joints are seen in the union of the bones of the skull. The edges of the bones are so fitted into each other that they form an unyielding joint, or suture.

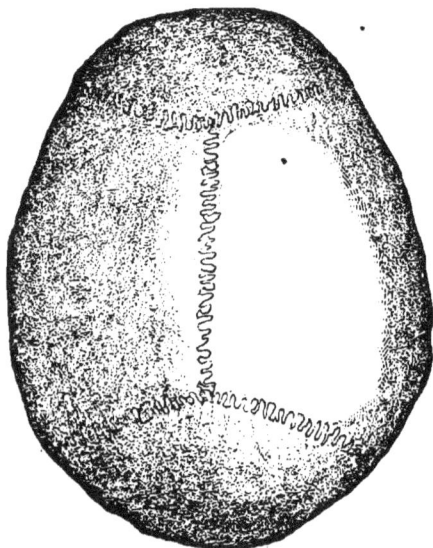

FIG. 68. The upper surface of the skull, showing the sutures, or immovable joints.

The Ball-and-Socket Joint. In the ball-and-socket joint, the head of one bone, which is more or less round like a ball, plays in the socket, or depression, of another bone. The bone with the round head can move in any direction, only the extent of its motion will depend upon the shape of the socket. If the socket be deep and small, as in the hip, the motion will be limited, but if the socket be shallow and broad, as in the shoulder, then the motion will be free in every direction.

Hinge Joints. The movements of the hinge joints are compared to those of a door. The elbow is the best example of a hinge joint. The movement is limited, for while the arm can be bent forward and then straight-

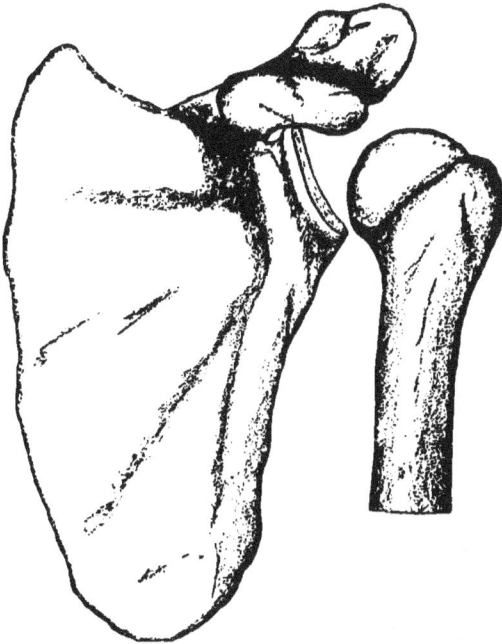

FIG. 69. The scapula and humerus in proper position to form the ball-and-socket joint of the shoulder.

ened, yet it cannot be bent backward. The knee is another example of this joint.

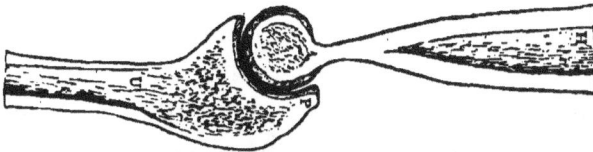

FIG. 70. A section through the elbow joint: H, the humerus ; U, the ulna ; P, the process, which prevents the arm from moving back of a certain line.

Pivot Joints. The rotary motions of the head, by which the skull and the atlas turn upon the pivot of the axis, furnish an illustration of a pivot joint.

SUGGESTIONS TO TEACHERS.

1. **The Marrow of Bone**. Cut a transverse section of a fresh bone; the bone in a round steak answers well. The oily matter, within the bone, is the marrow.

2. **The Periosteum**. Cut down on any fresh bone and peel off the membrane surrounding it. This is the periosteum.

3. **The Mineral Matter**. Burn a bone in the stove. The heat soon destroys the animal matter, leaving only the mineral. A burned bone is easily broken and pulverized with a hammer.

4. **The Animal Matter**. Place one of the long bones of a small animal, as the rib of a chicken, in a weak acid solution, — nitric acid, one part; water, four or five parts. In a few days pour off the acid, and place in water to wash out any remaining acid. If the bone has been in the acid long enough it can be bent and even tied in a knot. The acid has removed the mineral matter; the animal substance remains.

5. **Spongy Bone**. Saw any old dry bone lengthwise; especially the ends of any of the long bones. Notice the fine, honey-combed, spongy bone in the ends of the bones; while the body of the bone, between the ends, is hollow.

6. **Entrance of Blood Vessels**. Minute openings can be seen on the surface of any dried bone, as picked up in the field. Through these openings the blood vessels entered the bone, dividing again and again until small enough to enter the narrowest Haversian canal.

7. **Ligaments.** Any joint can be procured at the market and the outer layer of tissue removed. Still the joint is out of sight, covered with the hard, tough ligaments.

8. **Synovial Fluid.** If the joint be from a recently killed animal, when the knife cuts through the ligaments, a few drops of a watery substance escape. This is the lubricating synovial fluid.

9. **Joints.** With a saw divide the joint longitudinally, as represented in Figs. 66 and 70. Notice the working of the joint, the smooth articulating surfaces, and the great strength of the ligaments.

10. **Cartilage at the Joints.** In the last preparation, notice the cartilage covering the ends of the bones. Separate the joint completely and cut the cartilage with a knife. Notice the depth of the layer, and its shining, highly polished character.

11. **Cartilage between the Vertebræ.** Procure two adjoining vertebræ and cut them likewise. Examine the intervening cartilage. It appears elastic when pressed. It does not cut easily with a knife. This is not the same kind of cartilage as found at the joints.

12. **Connective Tissue.** When examining any piece of fresh meat, especially the outer part, soon after the skin is removed, a whitish tissue is seen. This can be pulled and stretched easily, with a pair of forceps. It is the tissue which holds, or connects the skin to the muscles and tissues beneath it. Connective tissue holds many of the tissues together and unites one kind of tissue with another. It is seen between the muscles, surrounding blood vessels, and forming the tendons to muscles. It is very generally distributed throughout the body.

CHAPTER XVII.

THE MUSCLES.

General Description. The great bulk of the body, external to the skeleton, is composed of muscles. The muscles give the general outline to the body and make nearly one half its total weight. Nearly all the muscles are arranged in pairs, so that the two sides of the body are almost alike. Some of the muscles are very small, as the minute muscles of the middle ear, while others reach from the hip to the knee. They are of a deep red color in the majority of animals, forming the lean meat or flesh. In many of the fishes, and in some of the birds the muscles are white, or of a light yellow color.

Two Kinds of Muscle. The muscles are, in respect to their function, divided into two classes; the voluntary, and the involuntary. The voluntary muscles are so-called because their movements are under the control of the will. Such muscles can be used whenever we wish or will to use them, as the muscles of the face or the arm. Others cannot be controlled in this way; they do their work independent of any action of the will, hence they are called involuntary. The muscles of the stomach and the heart are of this variety. The heart beats, the stomach contracts, and we are powerless to stop their action. As a rule, all those movements in the body most essential to life are not under the control of the will. Yet many of the involuntary muscles can be controlled, although for a short time only. An illustration of this is found in ordinary breathing. We breathe a certain number of times a minute and are entirely unconscious of it; still by an effort of the will, we can breathe faster, slower, or deeper. Even the voluntary muscles can be made to contract by a sharp blow, or by some fright. While voluntary muscles, therefore, are controlled by the will, they are not invariably so controlled. Nearly all the voluntary muscles are attached to bone at each end; while the involuntary are not attached to the skeleton, but are found in the walls of hollow organs, as the stomach and intestines, and in the walls of the arteries.

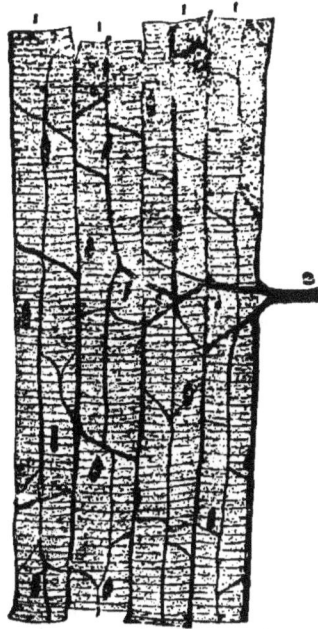

FIG. 71. Voluntary muscle, with its blood vessels: (1) the muscle fibers; (2) the blood vessels, magnified.

The Uses of Muscle. The muscles are primarily the organs of motion. They act as a protection to the blood vessels and nerves; they inclose the large thoracic and abdominal cavities; they serve as cushions to diminish the force of falls and blows; they fill up irregularities, and thus add to the symmetry of the whole body.

Structure of Voluntary Muscle. If a piece of lean meat,

which is voluntary muscle, be boiled, it will appear as if ready to fall apart into little bundles of tissue. These

bundles may be easily divided into still smaller ones, by separating them carefully with needles. In this way minute threads of tissue are obtained. If one of these be examined with a microscope it will be found to consist of many smaller threads, called muscular fibers. In Fig. 71, four of these fibers are seen side by side, with their accompanying blood vessels. In this figure and also in Fig. 72, fine lines are noticed running directly across each fiber. Because of these markings this variety of muscle has been called striated muscle.

FIG. 72. A muscular fiber, showing the nuclei, magnified.

On examining any piece of lean meat the bundles are seen as strings of red flesh, with white connective tis-

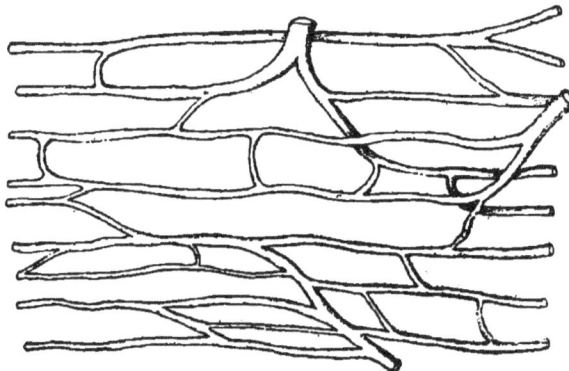

FIG. 73. The capillary blood vessels of muscle, magnified. The drawing is made from the same specimen as Fig. 71, only the muscular fibers are not shown.

sue between them. Boiling the meat dissolves this connective tissue to a certain extent, so that the bundles of fibers more readily fall apart.

Structure of Involuntary Muscle. Involuntary muscle is quite simple in its structure. It consists of a number of spindle-shaped cells, held together by a cement. This cement substance is found throughout the body. It is of the nature of glue, or cement, and it firmly holds many of the cells of the body together. Fig. 74, represents some involuntary muscle which has

FIG. 74. The cells of involuntary muscle, magnified.

been treated with dilute acid. The acid has dissolved the cement, and the cells are seen well separated from each other. These cells are very minute, and a high power of the microscope is required to see them.

The Tendons. The voluntary muscles are sometimes attached directly to the skin and to other soft tissues; but the great majority are connected to the bones by firm, white cords. These white, shining cords are called tendons. The tendons have no power of themselves to contract. They simply serve the purpose of cords, connecting the working part of the muscle with the part which it has to move. The parts acted upon may be removed a considerable distance from the body of the muscle; thus, the ends of the fingers are moved by the muscles of the forearm. The tendons serve another purpose: owing to their compact nature they occupy much less room than do the muscles, and thus the size of the wrist and ankle is much reduced. Were it not for this fact these joints would be covered with thick muscle, and it would be quite impossible to have the necessary freedom of motion. The tendons can be easily felt at the wrist, while the one attached to the thumb is easily seen on the back of the hand. The largest tendon in the body connects certain muscles on the back of the leg with the heel. Its attachment is shown at T, in Fig. 63. It is called the "tendo Achilles," from the Grecian fiction that it was at this point that Achilles received his death wound, as there was no other portion of his body that could be wounded.

Fig. 75 illustrates the muscles of the forearm, and their tendons. The tendons are held tightly down at the wrist by firm bands of tissue. Some of the tendons extend to the very ends of the fingers, so that when the muscles of the forearm contract, they move the most distant parts of the hand. The tendons are inclosed in sheaths, through which they easily glide.

FIG. 75. The muscles of the arm, ending in the white tendons at the wrist.

Muscular Contraction. By placing the fingers of one hand upon the fleshy part of the other hand at the base of the thumb (the ball of the thumb), the tissue will feel soft and comparatively thin. This tissue consists of voluntary muscle and can be made to contract by an effort of the will. With the fingers still in the position indicated, place the thumb on the end of the little finger; the muscle now feels thick and hard. From this we conclude that when a muscle contracts it becomes thicker and harder. We know a muscle shortens when it contracts, because it moves the parts to which it is attached.

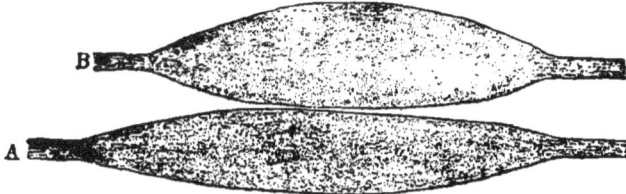

FIG. 76. A, a muscle relaxed, before it contracts: B, the same muscle, contracted; it is shorter and thicker.

A slight study of Figs. 77 and 78 will illustrate the principle upon which all voluntary muscles act. If the muscles on the front of the arm should shorten, the hand would be raised; while if the opposing muscles, on the back of the arm, should shorten, the hand would be drawn down again. If the muscles on the front of the leg should shorten, the toes would be raised; while the opposing muscles would raise the heel. Thus we learn that when a muscle contracts it becomes thicker, harder, and shorter; and that all the movements of the body are caused by such contractions.

In the case of the involuntary muscles, the individual spindle-shaped cells contract. If the cells be arranged in a circular manner, as around the arteries, then their contraction will diminish the size of the vessel. For the same reason, if the muscular cells that form the walls of the stomach contract, they will diminish the size of the cavity, and will force out its contents.

But a muscle cannot remain in a state of contraction for any great length of time. It soon tires and is

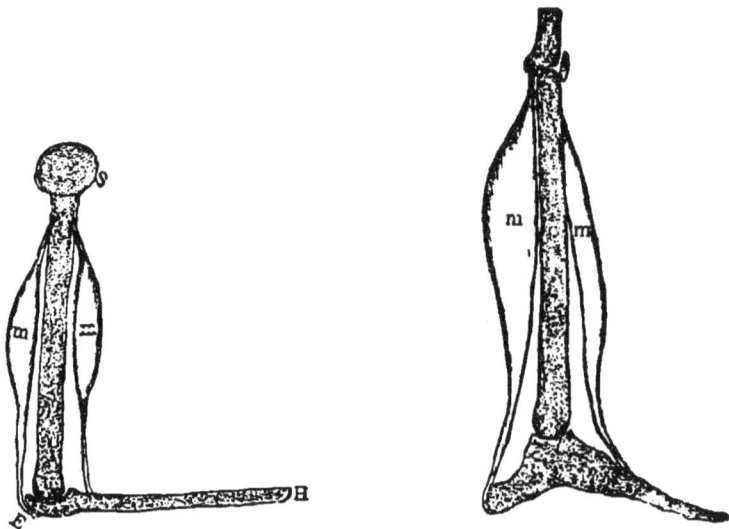

Diagrams illustrating the action of muscles.
FIG. 77. Muscles of the arm : s, the shoulder ; E, the elbow ; H, the hand ; m, muscles. FIG. 78. Muscles of the leg.

obliged to relax. After a short rest, however, it is again ready for work. All muscles must have rest or they will soon wear out. We learned on page 110 that even the most active of muscles, the heart, has a brief period of rest between its beats. We find that it is very tiresome to stand in one position for a long time, because, from continued tension, the muscles soon become fatigued, while a change in the position brings rest. For this reason it is much easier to continue walking, for a certain length of time, than to remain in one position equally as long.

Harmony in Muscular Action. The muscles which bend, or flex, the joints are called the flexors; while those which bring the bent parts back again, are called the extensors. Examples of the former are those muscles on the front of the arm which bend the forearm; while examples of the latter are those on the back of the arm which pull the bent arm straight again.

From this description it is evident that opposing muscles must not act at the same time; for if the flexors and extensors should equally contract, and pull upon the parts to which they are attached, there would be no motion whatever. To give free motion to a part, the opposing muscles must be finely adjusted to each other. In cases of spasms, or convulsions, the muscles do not act in harmony, and the body becomes stiff and rigid.

The delicacy of the adjustment of muscular action is well illustrated in the muscles of the face. Often the expressions of the face tell more than do uttered words. An unconscious contraction of a muscle, be it ever so slight, may convey, to the close observer, pain or pleasure. A slight contraction of a muscle lifts the brow, and a smile covers the face; while a change in another muscle is followed with a picture of suffering and pain. As a general rule, each muscle has a distinct motion to perform. It is also true that any one motion is usually brought about by the combined action of several muscles. The infinite variety of tones that can be produced by the human voice is due to the position and tension of the vocal cords, and these are controlled by muscular action.

The Control of the Muscles. All the muscles are under the control of the nervous system. The nerves serve as

a connecting medium between the brain and the spinal
cord, and the distant muscles. The nerve force is sent
from the brain, or spinal cord, down the nerve fibers to
the distant muscles stimulating them to contract. If
this connection be broken, there will be no muscular
action. It is possible, however, to make a muscle contract under such a condition, by applying the proper
kind of force from without.

Electricity might make a muscle contract when it had
no connection with the brain or spinal cord. It is even
possible to make a muscle contract, after it has been
removed from the body, by striking it a sharp blow or by
applying the electric current. But no force originating
within the nerve centers can reach the muscles, unless
the nerves leading to them be intact.

Standing. The ordinary act of standing, like so many
simple acts, is a result of the most varied and complicated conditions. Many opposite, or antagonistic,
sets of muscles contract and thus keep the joints rigid.
The flexors and extensors must equally contract; for if
one set should contract with more power than the other,
the body would surely fall. The young child finds
great difficulty in standing alone. The skeleton is not
only weak, but it appears quite impossible for the opposing sets of muscles to contract with equal power;
therefore the child has to think what he is doing, and
the task is a hard one. After the habit of performing
certain acts is formed, the muscles do their work without our being conscious of what they are doing.

QUESTIONS FOR CHAPTER XVI.

1. What purposes are served by the skeleton? Compare.
2. How many bones of the skull? How divided?
3. Describe one of the vertebræ.
4. Where is the atlas? What passes through its openings?
5. Where is the axis? Its peculiar feature?
6. How many ribs are there? How attached to the sternum?
7. Describe the bones of the upper extremity.
8. Describe those of the lower extremity.
9. Of what use is the arch of the foot?
10. How are the joints kept moist?
11. Of what use are ligaments? Injuries to them cause what?
12. Name the varieties of joints. Illustrate.
13. Where can each variety be found?

QUESTIONS FOR CHAPTER XVII.

1. Give a general description of the muscles.
2. Name the varieties of muscle. Why so called?
3. Give examples of each variety.
4. Give some of the uses of muscle.
5. Of what is voluntary muscle composed?
6. Describe the structure of involuntary muscle.
7. Where are tendons found?
8. Of what use are they?
9. How illustrate the contraction of muscles of the thumb?
10. This experiment shows what?
11. Why is it tiresome to stand long in one position?
12. How do we know a muscle shortens when it contracts?
13. Do all muscles require rest? Why?
14. What muscles are flexors? Extensors? Give examples.
15. The muscles are controlled by what?
16. What relation do the muscles bear to the nerves?

SUGGESTIONS TO TEACHERS.

1. **Voluntary Muscle.** Any lean meat represents voluntary muscle. Notice the bundles of fibers, and the white connective tissue extending around and through the muscle.

2. **Tendon.** A tendon is easily procured at the market. Notice how hard and unyielding it is. The muscles on the leg of a fowl show the tendons well.

3. **Action of Voluntary Muscle.** To illustrate the action of voluntary muscle, attached to bone, follow the directions on page 185, under the heading of "muscular contraction." For voluntary muscle not attached to bone, have the pupil pronounce the letter U; the voluntary muscle forming the fleshy part of the lips contracts. Follow this with the pronunciation of the letter Y, and the former muscles will relax.

4. **Involuntary Muscle.** The heart furnishes the best illustration of an involuntary muscle.

5. **Muscular Fibers.** Boiled corned beef shows the muscle, as if ready to separate into small bundles. Needles will tear the bundles into finer ones, while the microscope will show the smallest bundles composed of minute fibers.

6. **Expression.** A variety of experiments can be made to illustrate the fact that the various expressions of the face, as fear, courage, grief, joy, pleasure, displeasure, etc., are caused by the contraction of muscles.

CHAPTER XVIII.

EXERCISE.

The Necessity of Exercise. The blacksmith always has large arms, and the arm he uses the most will be the larger. This has been brought about by long and constant exercise of the muscles. It is only an instance showing that the exercise of a muscle makes it larger, harder, and stronger. If, for any reason, a person should be unable to use the arm for a few months, it would gradually become smaller and smaller; and if left unused long enough the muscles would nearly all disappear. This is illustrated in the following case: The arm of a young man had become completely paralyzed as a result of an injury, and so could not be used. All forms of medical treatment failed to restore any power of contraction to the muscles. In three years after the injury the person died of some acute trouble. A careful examination was made to find the muscles of the paralyzed arm. They had nearly all disappeared, but the small portions that remained were examined with the microscope. In place of the regular muscular fibers there were fibers nearly turned to fat, as represented in Fig. 79. Disuse of the fibers had caused them to undergo a fatty degeneration. From this and other facts, we conclude that exercise of the muscles is absolutely necessary to keep them healthy and strong.

But no one admires a man who is all muscle, and who has no brain; so we conclude that it is unwise to develop one particular part of the body and to neglect some other portion. The endeavor should be to develop all parts equally well. A proper amount of exercise is one of the essential conditions for the accomplishment of this end.

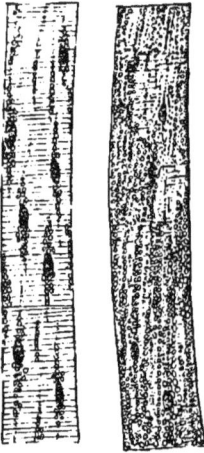

FIG. 79. Two muscular fibers undergoing a fatty change, magnified. The fiber to the left has a number of fat globules gathered near the nuclei. The changes in the fiber to the right are more advanced: the transverse lines have nearly disappeared; the nuclei are scarcely visible; and the whole muscle substance is nearly all changed to fat.

Many a young and enthusiastic scholar has been so carried away with his desire for mental advancement that every hour spent in other labor, or in rest, was regarded as so much time wasted. He failed to remember that the sound mind must be in a sound body to bring forth its highest and best results. It is to this class of devoted workers that we earnestly appeal. To all such we say, it is of vital importance that at least one or two hours of each day be spent in outdoor exercise.

General Exercise. Exercise should be taken out of doors as much as is possible, since pure air is of the greatest importance. No matter how cold the air may be outside, no one need fear taking cold if the body be kept in active exercise while exposed, and if rest be taken in the warm house.

General exercise should not be too violent; over-exercise is nearly as bad as no exercise. Any exercise is too violent which leaves the body "all tired out." Such exercise unfits the body for regular work, and may prove injurious to the nervous system. Healthful exercise should bring a restful feeling, a desire for work, and refreshing sleep.

Work and Rest. Each time a muscle contracts there is a waste of some of its substance. During the active work of a muscle, the waste far exceeds the repair. The worn-out material accumulates faster than it can be carried away, and the body experiences a sense of fatigue. If exercise be continued until the body is greatly fatigued, and if such exercise be frequently repeated, the muscles will gradually waste away as though they were not used at all.

If a muscle be made to work, it must have its periods of rest. The heart never appears to be tired. It beats on, year after year, with an astonishing regularity. But it would soon wear out did it not have a period of complete rest between its beats. In order that the muscles may be kept in a healthy condition there should be proper exercise followed by repose.

The Amount of Exercise. If exercise be so important for the general health, what is to be considered a proper amount? This varies within wide limits, according to the health and habits of the individual. If the organs and tissues of the body are poorly nourished, so that even slight exercise gives great fatigue, then the exercise should be very short, and followed by long rest. But the fatigue following the exercise of those in delicate health will grow less and less if the exercise be steadily continued ; always remembering to rest as soon as one begins to feel tired. On the other hand, a healthy body will exercise until all the muscles are thoroughly tired ; and yet, after a night of sound sleep will awake feeling all the better for the work.

Forms of Exercise. Walking is the gentlest form of active exercise. It throws into action nearly all the larger muscles of the body, except those of the arms. The advantage of this form of exercise, over gymnastics, is that it takes the person out of doors. Here the varied scenery has its exhilarating effect on the nervous system, and purer air is inhaled. To derive the full benefit of walking, it should be undertaken with a feeling of freedom and pleasure, and not simply because it has been prescribed by a physician ; otherwise, the exercise becomes so much hard work, and much of its benefit is thereby lost. Therefore it is better that this kind of exercise be given to the young in the form of games ; for that exercise is best which combines pleasure with muscular action.

Rowing is a very healthful exercise, and tends to develop many parts of the body. It is likely to prove injurious when long continued at a time, and when the muscles are used too violently. Horseback riding is an excellent exercise. It brings into play nearly all the muscles, while the fresh air and changing scenery impart a healthy tone to the entire nervous system.

Out-door exercise of some kind is always to be preferred to that taken indoors. How to find plenty of exercise in the open air does not trouble the farmer's boy, who has to get up early in the morning, do his part of the chores, and then walk a long way to school.

But he is the boy who eats heartily, sleeps well, and is not easily fatigued. He is laying the sure foundation for a healthy body. While it is true that any work which brings the muscles into play develops and strengthens them, yet it is equally true that "all work and no play makes Jack a dull boy." Therefore we enter a plea in favor of the ball and the racket for summer; and the sled, the skates, and the snowballs for winter.

Benefits of Exercise. The muscles are not the only parts benefited by exercise; the general health of the entire body is greatly promoted. Were this not true, but little would be said about muscular exercise; for simply to become physically strong should not be our highest ambition. "The pen is mightier than the sword." The mind of man is more to be admired than his muscular strength. It is because a healthy body is such a great aid to a vigorous mind, that an abundance of exercise is so persistently urged.

A proper amount of exercise increases the healthy action of the heart, and makes the blood flow more freely through the organs and tissues. It brings more air into the lungs, increases the appetite, and aids digestion. From this it logically follows that bodily exercise tends to give more activity to the mind, and to strengthen the mental powers. As the mind grows and expands it ought, under proper guidance, to bring forth all that is highest and best in man. Knowing these things, it is astonishing to us how any person can deliberately take into his system such poisons as alcohol and nicotine. He who indulges in these poisons is running the fearful risk of undermining not only the health of his body, but also the strength of his mental and moral nature.

When and How to Exercise. Vigorous exercise should never be taken either just before or just after a meal. A better time for exercise is when the stomach has about completed its work of digestion. A short walk before breakfast may give an appetite for the morning meal; but no very active work should be done until after food has been taken. A sudden increase in the amount of exercise is to be avoided. Excessive running, heavy lifting, and prolonged violent exercise of any kind are liable to be very injurious to those who have not been gradually brought up to such tests of strength. Running a race, or violently rowing a boat, by those who are unaccustomed to active exercise, might result in great harm. Exercise should be taken regularly, and then gradually increased. Never should attempts be made to far excel all former efforts. Let the body be gradually trained to withstand the severest tests ; then the heart and lungs will not suffer from over-exertion.

Physiognomy. Physiognomy is the art of discovering the ruling temper, or other qualities of the mind, by the external signs of the countenance. There seems to be some relation between the expressions of the countenance, and the qualities of the mind. Is there any physiological explanation of this relation ? It has been stated that the expressions of the face are caused by muscular contractions, and we already know that a muscle is strengthened by exercise ; therefore those muscles which are used the most constantly will be the strongest.

The explanation of the constant expression of the face is based on these facts. If the muscles that are used when we laugh are made to contract a great deal, and those used when we frown are exercised but little, then the former will become the stronger, while the latter will remain weak and undeveloped. This may become so marked that when a person is not thinking of laughing, the muscles still exert an influence, and there will remain a slight expression of laughter on the face. Such persons are said to wear a pleasant smile and a cheerful face. But suppose grief or pain has caused a person to cry a great deal; then there is left on the countenance an expression of pain or sadness. If one is in the habit of being cross and sullen, it will show in the expression of the face.

Thus, we conclude, the expression which is most constantly on the face is likely to become the permanent one. This is the reason why the face often tells so much of the true character. It is within the power of every person to conquer an irritable temper, and transform it into one of kindness and patience. This is not done simply by changing the expression on the face, but rather by filling the mind with true and noble thoughts; by cultivating those sentiments which are kind and charitable; and by promptly checking the uprisings of a quick temper. The expression is but the index of the force behind, a slight indication of the ruling thoughts and feelings.

ALCOHOL AND MUSCLE.

Perhaps no question connected with the study of alcohol is of more importance than the relation it bears to muscular activity. The man who is laboring hard, using his muscles many hours a day, wants to know if there is not something he can take which will tone up his tired muscles, and make them stronger. He desires to do more work and to do it more easily and more quickly. Many a laboring man thinks he cannot begin his day's work without a morning "tonic;" in the middle of the day he must have more, in order to go through the afternoon; while at night, more must be taken to drive away the tired feelings of the day. Is there an extra amount of work to be done? Then he resorts at once to some form of alcohol to carry him through the double labor.

The question is, *Does alcohol increase muscular strength?* The correct answer to this must be based upon carefully conducted scientific experiments, and confirmed by the widest observation among men. If alcohol does not increase the strength, why is it that so many declare it has this power? The workman in the shop says: "After taking a glass of beer, whiskey, or brandy I feel stronger and better able to work." Two questions must be considered as relating to this statement. First, is he actually made stronger, and second, are his feelings a true guide? We must admit that if alcohol will do no harm to any part of the body, and will impart strength to the muscles, it is greatly to be desired. To prove the correctness or incorrectness of the statement just quoted, it is necessary to study the effects of alcohol elsewhere. We cannot accept what he alone says, for he is already under the influence of the drug, and as we shall see later is therefore unable to decide impartially.

As a result of the most carefully conducted experiments, it is proved, beyond a doubt, that *both small and large doses of alcohol reduce the power of the muscles.*

Experiments have been made on soldiers who were given a fixed amount of work to accomplish. On certain days they were given some form of alcoholic drink, and on those days they were unable to work either so long or so well. Benjamin Franklin, when a poor printer boy, in London, was called the "American aquatic," because he drank only water, while his associates always drank beer. To prove to them that his "loaf of bread and pint of cold water" had more strength in them than their pint of beer, he carried a large form of type in each hand, up and down stairs, while his fellows could carry but one.

Experiments on individual men, and on companies of men, and extensive observation among men, prove, beyond doubt, that when the system is under the influence of alcohol, the muscles cannot exert their full power. The man who takes alcohol to increase his strength robs himself of muscular power. This may be a pleasant or an unpleasant fact ; nevertheless, it is one which science teaches and observation confirms. Four words contain the sum and substance of the whole matter: *Alcohol weakens the muscles.*

But why rely upon our statements alone for proof ? If we should make the statement that you could not possibly eat enough fruit or rich food of any kind to do you harm, you would reply that you know better. Your observation among your acquaintances is such that you could tell us of numerous instances where persons were made ill by over-eating fruit or indigestible food. Therefore, we say, look about you ; and draw your conclusions from your own observation.

Did you ever see an intoxicated man ? We hope not ;

but if you have, that one sad observation will make you decide the whole question. The manner of walking, the bent form, and the thick tongue, show the weakening of the muscles. For most positive and unmistakable proof, wait for the still further effect, when the whole body staggers, and, at last, falls to the ground. What is the result of this observation? Is this unfortunate man now stronger or weaker ?

Remember that the change from the full strength of a muscle, free from alcohol, to the weak muscle of the intoxicated man is one gradual and steady decline. Now, if the facts above stated have been carefully observed and considered, the conclusion is irresistible that one glass of an alcoholic beverage weakens the strength ; that two glasses diminish the strength still more ; that three bring it still lower ; and that with continued repetitions it must be seriously impaired or entirely overcome.

Finding that alcohol does not increase the muscular strength, we have the other question to answer : Why is it, then, so many people declare that alcoholic drinks make them *feel* stronger ?

We cannot deny their testimony ; yet how can we reconcile it with what we know to be the truth ? The whole trouble is just here ; alcohol is a narcotic, benumbing or diminishing the sensibility of the brain and nerves. The nerves should report to the brain the condition of the muscles ; but when the tired man takes an alcoholic liquor these reporters, the nerves, and their great center, the brain, are deadened and unable to give or receive an accurate report,— the inability varying in proportion to the amount of alcohol taken. This may be so marked that the person becomes quite

gay and light-hearted; the feeling of weariness or depression is gone because the nerves are deadened by
alcohol. But it is only the feeling that is gone. Let the
man test his strength and he can lift no more, indeed
not so much, and when the deadening effect of the narcotic has passed off he will feel all the more exhausted.

It will be a great day for America when all people understand the weakening power of alcohol. We believe there
are many thousands of men using strong drink, who
would stop it at once, if they only knew its power to
lower their ability to earn a livelihood. We live in a
country where the hardest and the best labor will surely
win. Do you wish to get to the top? You will need all
your powers, mental and physical. Alcohol will be as
an immense ball and chain fastened to your ankle while
you run to win the race.

We must remember another physiological fact: to
grow fleshy is not to grow strong. In athletic sports,
notice how the men who are to run in a foot race or to
row in a boat race, always train and diet for weeks in
order to get rid of their surplus fat, and to develop their
muscles. Fleshy people, as a rule, are neither so healthy
nor so strong as those who have less fat and more hard
muscle. An excessive amount of fat is one of the causes
of the "fatty heart" so common among beer-drinkers.
Much fat causes a disturbance in the action of the liver,
and is a hindrance to the proper action of the muscles.
Therefore, it is nothing in favor of beer that it makes
some persons fleshy. Beer never increases muscular
strength.

CHAPTER XIX.

THE SKIN.

General Description. The skin forms a strong, close-fitting garment, protecting the delicate and sensitive parts beneath. The skin is not fastened tightly to the tissues; it is held by delicate bands of tissue, which are of a loose or open nature. This allows the skin to be raised and gathered in folds. It also permits free movements of the skin over the joints and muscles. In some parts of the body, as the soles of the feet, and the palms of the hands, it acts as an elastic pad. It affords protection against external injuries, both from those of a mechanical nature, and from the action of chemicals or poisonous agents. In the case of a medium-sized man, the skin is equal to sixteen square feet of surface.

Adipose Tissue. The connective tissue beneath the skin usually contains more or less adipose, or fatty tissue. In thin persons there may be very little fat, and the outlines of the tendons and muscles, and even the shape of the bones may show through the skin. In fleshy persons the fat may be in great abundance, pushing out the skin, and causing the wrinkles and outlines of the parts beneath to disappear. In the average healthy body there is always some fat in the tissues beneath the skin. In old age the fat is likely to disappear, causing the skin to form in folds or wrinkles. The fatty tissue is of use as an aid in retaining the heat of the body, thus taking the place of so much extra clothing.

The fat is formed directly from the connective tissue. This tissue is composed of fibers and cells. Many of the cells are of the shape represented in Fig. 80. When the body is accumulating fat, small fatty or oily particles appear in these cells. Later these minute globules of fat increase in size and number until they run together as two drops of oil unite making one drop. Still

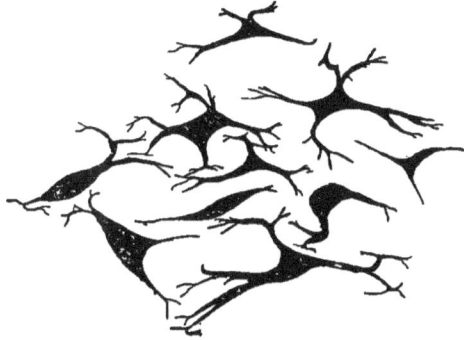

FIG. 80. Cells from connective tissue, magnified.

later the entire cell becomes filled with the fat, giving the appearance of one large spherical body, holding a large fat globule. The connective tissue-cells thus become greatly distended and thereby increased in size. When the fat in the

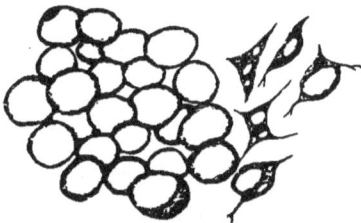

FIG. 81. Fat cells, magnified: to the right are five connective tissue-cells partly filled with fat.

body is being reduced by disease or starvation, it gradually disappears from the cells.

The Epidermis. The skin is composed of two layers, the epidermis and the dermis. The epidermis, which is also called the cuticle, or false skin, forms the outer layer. It is composed entirely of cells, the outer ones being very hard and dry. The cells are arranged many layers deep ; through them pass the ducts of the sweat glands, the oil glands, and the hair shafts. Fig. 82 shows

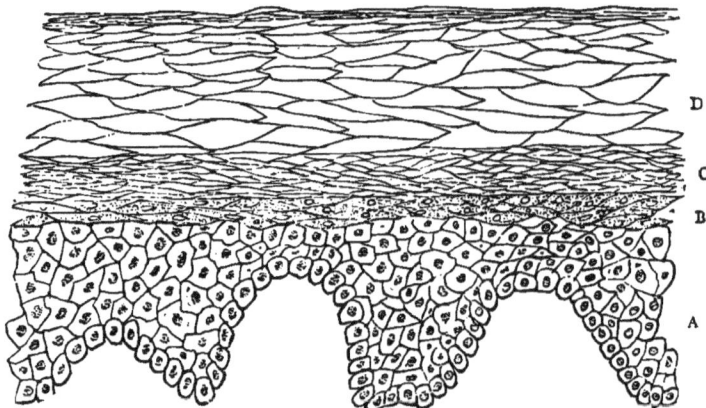

FIG. 82. A section of the epidermis, magnified : A, the layer of cells nearest the dermis ; D, the layer nearest the surface of the body.

the microscopical structure of the epidermis. The outer cells at D have no nuclei, are very hard and dry, and are being constantly removed as mentioned below. The epidermis has neither blood vessels nor nerves ; it is, therefore, bloodless and without feeling. For these reasons it is possible to remove nearly all the epidermis by gently scraping the skin with a knife without causing pain or the flow of blood. But as soon as the true skin, the dermis, is reached, the scraping brings both pain and blood.

The epidermis is removed entire from the dermis by a blister. An ordinary blister on the hand shows a thin membrane, raised from the parts beneath, and separated from them by a watery fluid. Such a membrane may be cut without pain or blood ; for it is the epidermis of the skin. After the thin membrane is removed, if the red surface beneath it be touched it will be found highly sensitive and easily made to bleed.

Epidermis Rapidly Changing. The epidermis is a good illustration of the wear and waste of the body. The outer cells are constantly falling off in vast numbers ; immense numbers are removed daily by the friction of the clothing, and by the work of the sponge and towel at the bath. This great loss is being steadily made up by the formation of new cells in the deeper parts ; these come to the surface as the ones over them disappear.

If all of the epidermis were removed at once the parts left would be very red and tender; but in the case of the blister, a few cells of the epidermis are left clinging to the true skin; these rapidly multiply until within a few days a complete new epidermis is formed. After some fevers, as during the " peeling" of scarlet fever, large masses of cells are removed together.

The Coloring of the Skin. The color of the skin depends largely upon the character of the deepest cells of the epidermis, as at A, Fig. 82. In very light skins these cells are colorless, in darker skins the cells have a slight amount of dark coloring matter in them ; while in the darkest skins, the coloring matter is very abundant. A destruction of these deep cells causes the epidermis to appear perfectly white, as it does in certain diseases. The white skin of the Albinos is due to the absence of any coloring matter in these cells. A free supply of blood to the skin gives it a red or pink color ; while any interference with the action of the liver may give a jaundiced, or yellow color to the skin.

Uses of Epidermis. The outer layer of cells of the epidermis, as seen at D, Fig. 82, consists of closely packed, hard cells which make a very complete and almost impenetrable covering. Therefore, the epidermis is a protection against the absorption of poisons. The physician understands this; for if he wishes to introduce medicine into the system by the skin, he first scrapes off the epidermis, or removes it with a blister, and then sprinkles on the drug. In vaccinating, it is necessary to remove the epidermis, in order that the virus may come in contact with the absorbent vessels beneath.

The surgeon is not harmed while operating on diseased portions of the body, because the epidermis prevents the absorption of any decaying matter; but a few cases are on record where the skin on the surgeon's hand was accidentally cut or broken while operating, thus allowing diseased matter to be absorbed, producing blood poisoning and even death. Immersing the whole hand in poisonous matter might possibly do no harm, provided the epidermis were in a perfect condition everywhere; but the slightest prick of the finest needle might then be the cause of death. The epidermis also protects the parts beneath from the sudden changes of heat and cold.

The Tactile Bodies. The tactile bodies give the sense of touch. They are situated in the dermis, or true skin; they reach to the epidermis but do not penetrate it. Wherever the sense of touch is the most delicate, there are found the largest number of tactile bodies. Over one hundred of these bodies have been counted in a space $\frac{1}{50}$ of an inch square. They are very small, averaging no larger than $\frac{1}{350}$ of an inch in length. Thus the microscope shows that while it is true that the epidermis has neither blood vessels nor nerves, yet just beneath it are nerves especially arranged for the sense of touch. These tactile bodies are illustrated at 4, Fig. 83. The dermis is also well supplied with blood vessels.

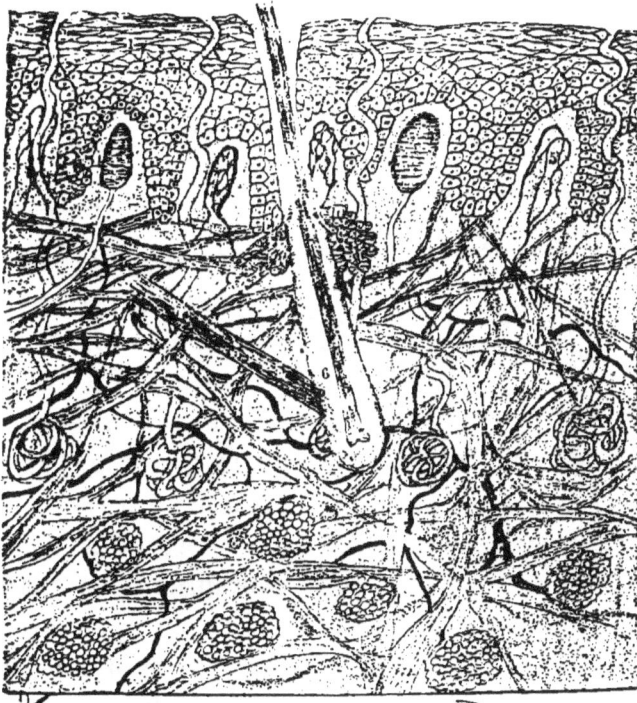

FIG. 83. A section of the human skin, magnified : (1) the epidermis ; (2) the duct of a sweat gland ; (3) a sweat gland ; (4) the ending of a nerve, for the sense of touch ; (5) coils of minute blood vessels ; (6) a hair follicle, in which is a hair ; (7) a muscle, which can move the hair follicle.

The Sweat Glands. The sweat glands are found in the deep parts of the dermis, or in the tissue immediately beneath the dermis. In Figs. 83 and 84 these glands are represented as minute tubes arranged in circular coils. That part of the tube which is not coiled, but which extends from the gland to the surface of the skin.

is called the duct: the duct pursues a spiral course, as represented in Fig. 84, and opens on the surface with a funnel-shaped dilatation.

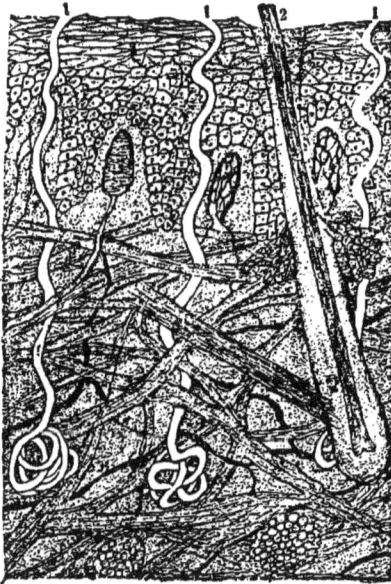

FIG. 84. A portion of Fig. 83. The section was prepared especially to show the sweat glands and their ducts : (1) ducts of the sweat glands ; (2) the hair shaft.

If the ridges which appear so plainly on the ends of the fingers and palms of the hands, be examined with a small magnifying glass, it is possible to see the minute depressions in the center of them ; these represent the openings of the sweat glands. The view obtained will resemble that given in Fig. 85. The openings are quite close together, in some places averaging as many as 3000 to the square inch. From the surface of the skin to the coiled glands is about one fourth of an inch. It is estimated that there are over three millions of these glands in the entire skin.

The Perspiration. The perspiration is a colorless fluid, secreted in the coils of the sweat glands. It is of very simple composition, over 99.5 per cent of it being water. Some inorganic substances, as sodic chloride, common salt, give it a salty taste; while some organic ingredients, as the fatty acids, impart to it an odor. The perspiration is a continuous secretion. When it is small in amount, the water is evaporated from the skin at once, therefore its presence on the skin is not noticed; this is called the insensible perspiration. As soon as the secretion is increased, it does not all evaporate, but gathers as drops of sweat on the surface; this is called the sensible perspiration. Many conditions cause the amount of perspiration to vary. In some individuals the whole amount per day is very small; while in others it

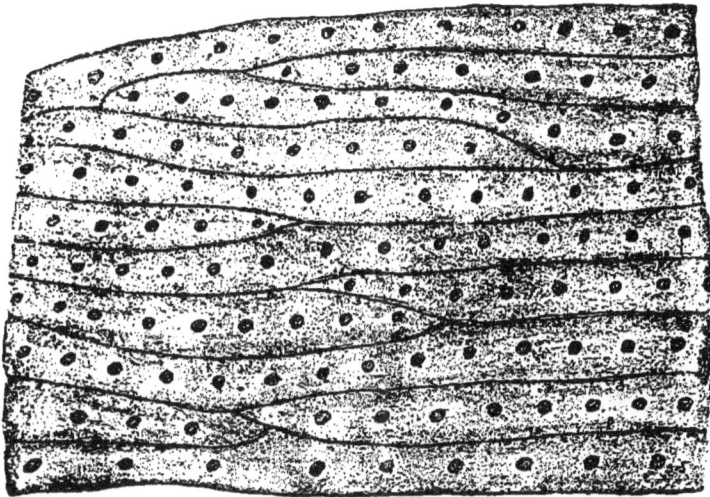

FIG. 85. The surface of the skin, slightly magnified, showing the openings of the sweat glands.

is very large. It is a fair average for the year through to say that from two to four pounds are secreted each day.

There is a marked difference in the perspiration of the lower animals; the horse perspires freely; the ox to a slight extent only ; and the dog but little, if any. The panting of the dog after exercising allows much water to be given off from the body through the lungs;

14

and in this way the object of regulating the temperature of the body is gained, as will be described later.

Controlled by Nerves. The activity of the sweat glands is under the control of certain nerves. While it is true that dilatation of the blood vessels is usually associated with profuse perspiration, yet it is possible to arouse these glands into activity even after the supply of blood has been cut off. For instance, in cases of great fear, when the skin is extremely pale, and its blood vessels nearly empty, the face may be covered with great drops of perspiration.

Conditions Affecting Perspiration. An increase in the temperature of the surrounding air will cause an increased secretion of the perspiration. Individuals who work in furnaces, or in places of very high temperature, often perspire as much in an hour as the average person will in a day. Those who perspire freely ought to drink a large amount of water to supply the loss, or the system will soon become exhausted. An extra amount of water in the system, as there is after a free use of warm drinks, will increase the secretion. Muscular activity is known by all to do the same. Certain drugs excite very copious perspiration, while other drugs diminish it. The secretion is lessened by cold; in fact, it may be suddenly arrested by this means.

The liability to perspire varies greatly in different individuals, and in the same individual at different times. Any departure from health may cause it to vary from the hot, dry skin of fever, to the profuse night-sweats of consumption. There is a direct relation existing between the kidneys and the skin. In summer when the skin is active the secretion of the kidneys is much lessened, while in winter when the skin is less active the work of the kidneys is increased.

Object of the Perspiration. The chief object of the perspiration is to regulate the temperature of the body, although it removes a slight amount of worn-out material. It is well known that when a liquid evaporates it produces cold. So the evaporation of a number of pounds of water each day from the surface of the body during the warm days of summer, causes a considerable lowering of the temperature. Exercise during the summer would be quite impossible were it not for this fact. The exercise causes rapid oxidation of the tissues, and this produces heat; and added to this is the heated atmosphere. Were not some way provided of cooling the body, it would soon be in a raging fever. But the more exercise, and the higher the thermometer, so much more profuse is the perspiration, while its rapid evaporation causes the body to remain at about a fixed degree of heat.

Checking the Perspiration. Sudden cooling of the skin checks its action, throws additional work on other parts of the body, and often causes disease. One of the most frequent causes of a cold is the sudden checking of the perspiration. After exercising, or whenever the body is perspiring freely, there should be great care in regard to draughts of air. The body should be gradually cooled, with some light clothing thrown over the shoulders, while resting. Sudden checking of the perspiration is positively injurious and may easily lead to fatal results.

The Hair. A hair consists of the root and the shaft. The former is situated in the skin, and the latter projects from it. The hair may be easily removed from its sac or follicle, without damage. At the lower end of each hair is a small eminence, or papilla, which is well supplied with blood. The root of the hair rests on this papilla and grows therefrom. The cells of the papilla multiply and grow, pushing those already formed upwards toward the surface. A hair therefore, grows entirely from the root, from this minute papilla. If a hair be removed, another begins to grow at once from the papilla, and in time will appear on the surface. A destruction of the papilla is necessary to prevent the hair from growing.

The hair shaft is not hollow, although its center is composed of cells more loosely arranged than those forming the exterior. Hair is very elastic; with proper care it can be made to stretch nearly one third its entire length before breaking. It is also very strong, — a single hair being capable of suspending a body weighing from three to five pounds.

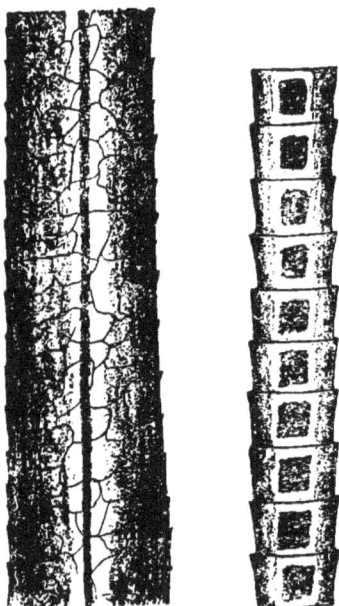

FIG. 86. Human hair, magnified.   FIG. 87. A hair from the cat.

A glance at Figs. 86 and 87 shows that the microscope reveals a great difference between the hairs of the lower animals and those of man. In certain cases the difference is very marked, as here shown, while in others it is not so marked. These facts have been used for the detection of fraud, and other crimes.

Muscle of the Hair. In Fig. 83 it is noticed that the hair is placed obliquely in the skin. Fastened to the

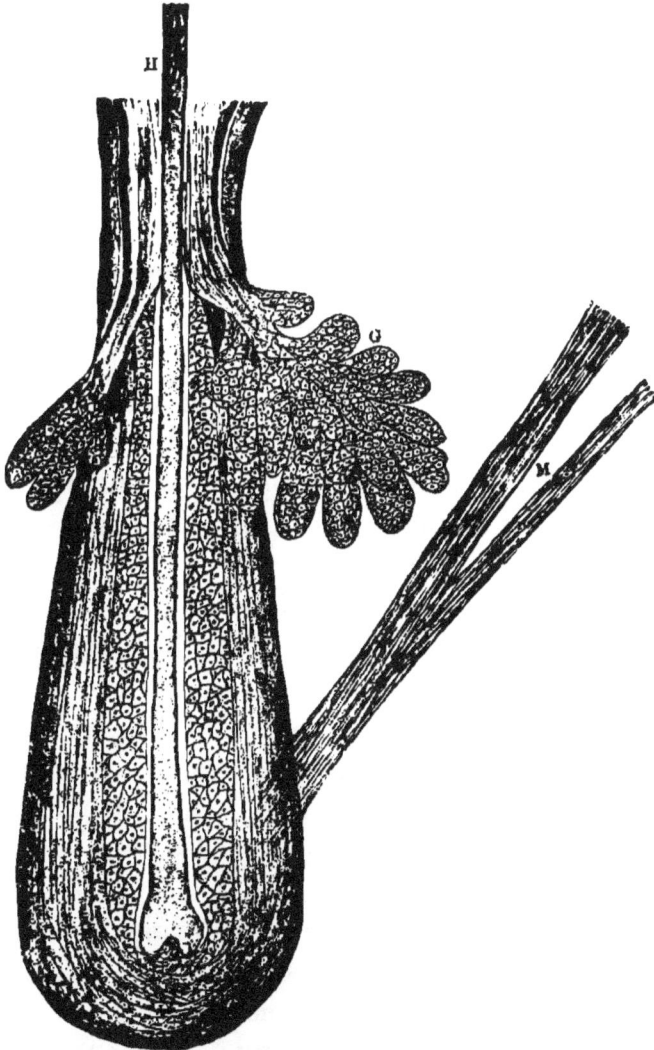

FIG. 88. A human hair in its sheath, or follicle, magnified: H, the hair shaft; M, the muscle; G, the oil gland.

lower part of this hair sac and extending obliquely upwards and to the left is a muscle. This muscle is better shown at M, in Fig. 88. The muscle is thus arranged at an acute angle, so that when it contracts, it pulls on the base of the hair sac, causing it to stand more nearly erect. In this way the hair is made literally "to stand on end." A contraction of this muscle produces the condition known as "goose skin."

The Sebaceous Glands. The sebaceous or oil glands are situated by the side of the hair sacs into which they open by a duct, as illustrated in Fig. 88 at G. These glands secrete an oily substance which is spread around the hair making it smooth and glossy. Some of the secretion extends over the skin making it soft and thus preventing it from becoming hard and dry. Often the ducts to these glands become stopped up, and the secretion distends the glands. This afterwards becomes dark colored, disfiguring the skin.

By referring to Fig. 88, it will be noticed that a sebaceous gland is situated between the hair sac and the muscle, M. By the contraction of this muscle the hair is drawn toward its upper attachment and the gland thereby compressed, thus aiding in forcing out its contents.

The Nails. The nails grow from behind forward, thus being constantly pushed outward. They are composed of minute cells, similar to those found in the deeper parts of the epidermis, only they are firmer and harder. The nails protect the ends of the fingers and toes; and they give aid to the fingers in picking up small objects. If a nail be removed by an accident, a new one will take its place in a few weeks, providing the root is not injured.

CHAPTER XX.

BATHING. — CLOTHING.

Necessity of Bathing. We know that an immense number of sweat glands are constantly pouring their secretion on the surface of the skin ; that vast numbers of oil glands are depositing an oily substance on the surface also ; and that the cells of the epidermis are constantly loosening and falling off in great numbers. It is only necessary to recall these facts, in order to understand fully the necessity of frequent bathing. The glands of the skin have a certain work to do ; how can they properly perform this if their ducts be closed by an accumulation of such material on the surface ? If these glands cannot perform their full duty, then other glands or organs must do it for them, or sickness will follow. We conclude, therefore, that neglect of the skin means more work for the lungs and kidneys, especially for the latter.

Always considered Important. The nations of antiquity had much to say about the bath, both as a luxury and as a means of preserving the health. The public baths of Rome were among the most interesting of her works of grandeur and beauty. Most beautiful works of art have been recovered from their ruins, all speaking of the splendid preparations made for this luxury. The cost of a bath in one of these elegant quarters was almost nothing, so that all people could enjoy its benefits. After bathing, the skin was usually anointed with perfumed oil; and this was followed by light exercise for a short time. In those days nearly every one could swim; and to be unable to swim was about as great a disgrace as to be unable to read.

When to Bathe. Probably immediately after rising in the morning is the best time to bathe. The body is rested, reaction is easy, and the circulation is at its best. The ideal bath is a brief, daily use of cold water, immediately after rising, followed with a brisk use of the towel. One should never bathe when greatly fatigued, nor use a cold bath if feeling chilly. Neither should a bath be taken immediately before or immediately after a hearty meal. It should not be taken soon after a meal, because the rubbing brings the blood to the surface, and therefore must take some of the blood away from the stomach, where it is needed. Great caution should be used in bathing when the body is over-heated, and when it is perspiring freely; because if the water should be too cool, or if the rubbing should fail to produce a complete reaction, then a severe cold or something more serious might result. Both too cold and too hot water are alike injurious.

The Cold Bath a Tonic. A healthy person should regard the bath as something more than a cleansing process. It should be a tonic of the most invigorating character. The first effect of the cold is to drive the blood from the skin; but soon reaction takes place, and the blood returns with renewed force, filling the capillaries of the skin, and imparting a healthy glow to the entire surface. Just at this time the bathing should cease. The brisk use of the towel heightens the flow of blood, and the whole body becomes enveloped with a pleasant sense of comfort and warmth. To remain longer in the bath would be to send the blood from the surface for the second time, from which reaction might not occur, causing a sensation of chilliness and fatigue. The brisk work of the rubbing· causes the exercise of many muscles, and the whole body is thereby aroused to activity.

The cold bath is injurious to those who do not have a quick reaction from its use. The skin should be left red and warm, and the whole system ought to be invigorated. If not, and the opposite be true, then the cold bath is injurious rather than helpful. If the bath be short at first, if the water be cool, if it be continued daily, if the towel be used briskly, and if only a portion of the body be bathed at a time, then it must be of benefit. Soon the water can be used colder and the time of the bath extended. With these precautions in mind, it is safe to assert that there are very few young people, in fair health, who would not be greatly benefited by the daily use of the cold bath. A quick sponge bath with cold water, followed by thorough rubbing, is the best method to illustrate the advantages and safety of the cold bath.

Salt Water Bathing. Many persons who visit the seashore plunge into the cold water at once, and remain there for a number of minutes. If they have been accustomed to the cold bath at home, there can be no more healthful exercise. The beating of the surf against the skin is an invigorating tonic in itself, while the muscles of the whole body are brought into action in withstanding the force of the waves. The change of scenery and the freedom from care aid in imparting new life to the seaside resorter. If he be unaccustomed to the cold bath, however, and suddenly spends from twenty to thirty minutes in the sea, he must expect undesirable effects to follow. A vast amount of sickness is caused in this way, all of which might be prevented. A person should not remain over three minutes in the water at first, and then should immediately rub the body briskly with a towel, and dress warmly. To such as follow this advice sea bathing, in most cases, will be of the greatest benefit. The best time for a bath in the sea is before the midday lunch, about 11 A. M. This gives time for a nap or a short rest before taking food into the stomach. It is a great mistake to suppose that one needs to take an alcoholic drink of any kind after bathing to prevent taking cold.

Warm Baths. An occasional warm bath, with the use of pure soap, is necessary. But warm water should not be used for the daily bath, unless advised by a physician for the purpose of removing impurities from the system through the skin. In such cases the steam bath, the Turkish bath, or the Russian bath is often prescribed. A daily warm bath lowers the tone of a healthy system and diminishes its power to withstand disease.

How to Bathe. There is always a temptation to have the bathing-tub filled with tepid water, and then remain in it a long time. But this method deprives the bath of its tonic properties. A bath of this description once a week may be desirable, but is most undesirable for daily use. The better method is to use the hands, or mittens made from crash toweling; quickly cover a portion of the body with cool or cold water; rub this portion dry; and thus proceed until the bath is completed. The whole bath should not exceed five minutes.

The Value of the Bath. The value of the bath will depend largely on the completeness of the reaction following its use. As stated before, the skin should be warm and the whole system refreshed. But if the skin be cold, and the body feel chilly and fatigued; or if the bath be followed by headache and general lassitude, then the practice should be discontinued, and a physician consulted to discover where the error rests. The bath is of great value to those who are already well and strong; it is to such that its use is so freely recommended. It is of great value in keeping the various organs and tissues of the body in a healthy condition, and it is therefore a good preventive of disease. The aged and the feeble need special advice for their particular cases; and the bath should be used by them only under the advice of a physician.

A Good Complexion. Bathing is essential if one desires a good complexion. The skin must be active and do its part of the work. Frequent bathing of the face with pure water is especially desirable. Soaps are not necessary for the face, while their use often causes roughness of the skin. Cosmetics almost invariably contain substances which are injurious to the skin. Their continued use often seriously affects the general health.

The Clothing. One object of the clothing is to prevent too great loss of heat from the body. The surrounding atmosphere is nearly always cooler than the body, therefore there is a tendency for the body to lose heat. The clothing corrects this tendency in a large measure. Food produces the heat and the clothing prevents its escape; for this reason it is true that poorly fed persons need more clothing in winter than those who are well fed. Animals require less food, and are able to do more work, if they are kept warm during the cold weather. Protection from the cold economizes the fuel of the body, which is the food. Clothing protects the body from external injuries, and is also an ornament; but its great object is to aid in maintaining an equal temperature in the body.

Nature often Regulates the Clothing. The lower animals are unable to alter their clothing to meet the change of the seasons; so Nature does the work for them. In winter the long fur and the heavy coat of hair protect the body from the wind and snow; in the spring this is changed, and a thin, light covering takes its place. Whether this covering be thick or thin, of fur or feathers, it is always a poor conductor of heat. A most valuable lesson may be learned from this fact: no matter what the outer garments may be, those next to the skin should be made of a material that is a poor conductor of heat.

The Materials of Clothing. With the above facts in mind, it is natural that furs should be extensively used during the cold weather. Next to fur in this respect is wool. All woolen garments are poor conductors of heat, and are therefore valuable for winter use. Woolen garments are a wonderful protection against the cold in winter, and during the sudden changes in summer, they should be worn next to the skin the entire year.

The thick and heavy material can be used during the winter, and the thinnest and lightest during the summer. If woolen garments irritate the skin, then silk may be used, as approaching the nearest to wool. These simple precautions in dress will prevent an immense amount of illness. The sudden changes in temperature and exposure to draughts would not so often be followed by colds, chills, sore throats, and lung troubles, if the body were constantly surrounded by a non-conducting medium, as wool or silk. Linen is a good conductor of heat; therefore, it should not be used next to the body. Cotton is much to be preferred to linen, and it is much better for bed-clothing. The outer garments can be regulated according to the seasons and the conditions of the air, but under all circumstances keep the surface of the skin well protected with either wool or silk.

The Weight of Clothing. A heavy covering of cotton, or linen, is not so warm as a lighter covering of wool or silk. The weight of clothing has no definite proportion to its warmth. Air is a poor conductor of heat, so is wool; hence, it follows, that the light, loose cloud worn by ladies as a covering to the head is based upon scientific principles. The air in the meshes of the wool, and the wool itself, give little chance for the heat to escape. Therefore in choosing clothing, the nature of the material should be considered more than its weight, — its value as a non-conductor of heat being in the order here given : wool, silk, cotton, and linen.

Clothing should Fit well. It is not conducive to the general appearance to have ill-fitting clothing; yet it does not follow that tight clothing adds to the appearance. To the physiologist, the constricted waist and the pinched foot represent ill-fitting clothing decidedly more than does the loosest hanging garment. Tight belts and corsets are accountable for a long series of complaints. Constricting the waist compresses the liver and stomach and interferes with the movements of the diaphragm ; this often causes headache, dyspepsia, shortness of breath, and a multitude of aches and pains. Clothing can and should be made to fit well, and yet it need not interfere with the action of any organ, or with the natural movements of any part.

Clothing should be Changed. Wet feet are the cause of many a sore throat and severe cold ; and for most persons, it is a great risk to allow them to remain damp. They should be dried and rubbed thoroughly as soon as possible, in order to fully restore the circulation. If one be caught in a storm so that the clothing becomes damp or wet, it should be changed at the first opportunity. Brisk exercise in the mean time will keep the body from becoming chilled. But if chilly sensations and hot flashes are already creeping over the body, probably a severe cold is coming on.

To Cure a Cold. A cold is generally produced, as just indicated, by some exposure, so that the skin is inactive and the internal organs congested. The mucous membrane of the nose, throat, larynx, bronchi, and the lungs are, one or all, most likely to be affected. During the summer the congestion of the internal organs often causes a disturbance of the digestive organs. It is essential, therefore, to cure a cold at the beginning; this can often be done by very simple remedies. The object to be accomplished is to restore activity and warmth to the skin. This can be brought about by giving hot drinks, and by adding extra clothing. Drink slowly a bowlful of hot lemonade, or hot ginger-tea, while the feet are in a hot bath, and while an extra blanket is thrown over the body. Soon the body will perspire freely, and the whole system feel warm from the artificial heat. Have the bed-clothing thoroughly warmed before getting into bed, and then keep well under cover. Remain in this condition until the body is thoroughly covered with perspiration, when we can be assured the skin has resumed its activity. Have the body rubbed dry, and gradually remove the extra clothing, in order that the temperature may be slowly lowered. If such simple treatment were more frequently and more quickly carried out, a vast amount of illness would be prevented.

QUESTIONS.

1. Give some reasons why frequent bathing is necessary.
2. When is the best time to bathe ?
3. Give some cautions about bathing.
4. Describe the first effect and the reaction of a cold bath.
5. When is a cold bath injurious?
6. Give some directions about sea bathing.
7. What is said about the warm bath ?
8. The value of a bath depends upon what ?
9. Why is the bath a good preventive of disease ?
10. What is one object of the clothing ? Explain.
11. How does nature regulate the clothing of the lower animals ?
12. What material is a poor conductor of heat ?
13. Is heavy clothing necessarily the warmest ?
14. Give directions for curing a cold.

* CHAPTER XXI.

ANIMAL HEAT.

Sources of Animal Heat. It has already been stated that the heat of the body is derived from the food, and from the oxygen obtained during respiration. By the union of these, oxidation occurs, and heat is produced. It appears from this statement that there must exist a relation between the amount of oxygen consumed and the amount of heat produced in the body: thus, if an animal consumes little oxygen it will have a low temperature ; if much oxygen, then a higher temperature. Another source of heat is from certain physical processes of the body, as the work of the heart, the general circulation, and the active exercise of the body.

The chief source of heat is found in the muscular system. The muscles form a large proportion of the whole frame, and they are very active during many hours of the day. The greater their activity, the more rapidly will the tissue be exhausted, and new tissue take its place. These changes require the oxidation of much food, and thereby much heat is developed. Next to the muscles are the secreting glands. Most rapid changes occur in these glands when they are active, all of which produce heat. The liver is the most important gland in producing heat. The changes taking place in the liver cells are very active and continuous. The warmest blood in the body is found just as it leaves the liver, on its way to the heart, being much warmer here than when it enters the 'liver. But heat is generated in every organ and tissue in the body; as each activity contributes to an elevation of the temperature.

Cold-Blooded and Warm-Blooded Animals. Cold-blooded animals are those whose temperature is generally about the same as that of the air or water surrounding them. They consume little oxygen, and therefore a small amount of heat is developed. Frogs, reptiles, and fishes have a low temperature, which varies at times between wide limits. When the temperature of their surroundings is increased they consume more oxygen, and thus develop more heat; while if the temperature be lowered the amount of oxygen consumed is diminished, and the animal heat reduced.

During the winter the frog buries himself in the cold sand at the bottom of the water, where he receives enough oxygen through the pores of the skin to sustain life. But during the warmer weather of summer he comes to the surface to fill his lungs with air; this is necessary in order to meet the demands of the more rapid changes in the tissues of his body. During the winter his home is under the water, but during the summer he would be drowned if he remained in water, as the more rapid oxidations require more oxygen than can be supplied through the medium of the skin.

Warm-blooded animals are those whose temperature is generally above that of the surrounding air. They maintain a remarkably uniform temperature throughout the heat of summer and the cold of winter. A person may ride a long distance in the cold air, or work hard in an over-heated room, yet the temperature of his body will remain almost at a fixed point.

Temperature of the Body. The thermometer shows that different parts of the body vary in temperature. In those parts where rapid changes are taking place, and where oxidation is most marked, the temperature is much higher than the average for the body. The blood is constantly passing from one tissue to another, carrying warmth from the tissues where heat is being developed to other tissues where it is being lost; thus the blood tends to equalize the temperature of all parts of the body. The temperature is ascertained by placing the bulb of a thermometer under the person's tongue.

The loss and production of heat are so evenly balanced that the temperature of the healthy adult body varies little from $98\frac{1}{2}°$ F. This is maintained with only slight variation throughout life. So accurately is this adjusted during health that a variation of more than a single degree denotes some disturbance in the system; a fall of two degrees below the normal temperature is considered a serious matter; while a severe cold may cause it to rise two degrees above normal. A temperature above 103° denotes a high fever; of 105° a severe attack; above 105° is most alarming; while recovery after the thermometer has recorded 110° is very rare.

The normal temperature of $98\frac{1}{2}°$ F. is subject to some variations within narrow limits. There are quite regular variations in the course of every twenty-four hours. The temperature continues to rise during the day until it reaches the highest point from five to eight in the evening; then it continues to fall during the night until from two to six in the morning, when it is at the lowest. About the middle of the forenoon, or about three hours after the morning meal, the thermometer should record 98½° F. The difference between the lowest and highest points reached during the day probably does not exceed one degree.

The Regulation of Heat. The question now arises, how is the excess of heat above 98½° removed from the body ? The body would soon become very much warmer were not some means provided for regulating the heat. It is estimated an adult body produces, in one hour, enough heat to raise the temperature three degrees. If no heat were given off, in thirty-six hours it would reach the boiling point ; hence the distribution and removal of the excess of heat becomes an important matter.

The two principal tissues which regulate the temperature are the lungs and the skin. It has been stated that the expired air is warmer than the inspired air. Therefore considerable heat is required to produce this warmth, which is carried out of the body with each expiration. The evaporation of the water of the expired air — the watery vapor — also produces cold ; indeed the loss of heat is in definite proportion to the amount of air taken into the lungs in a given time. The rapid breathing of a dog after running, and the extra flow of water from his mouth are illustrations of these facts.

But the work of the skin is far more important, being five times as great as that of the lungs, in this particular. It must be evident that the more freely the blood passes through the skin, thus coming under the influence of the cooler surroundings of the body, so much the greater will be the loss of heat. The evaporation of the perspiration results also in a great loss of heat to the body.

This regulation of the heat is well illustrated by studying the changes which take place during some active exercise. Muscular contraction gives rise to heat; hence exercise must increase the bodily temperature. But the thermometer shows no such change; what then becomes of the heat thus produced? The exercise causes rapid breathing, and hence more loss of heat through the lungs; while more blood is sent to the skin, where its temperature is lowered. The exercise also causes the skin to perspire freely, and the evaporation of the water from the surface of the body results in a great loss of heat. Thus we find that the extra amount of heat caused by muscular exertion is completely lost by the increased action of the lungs and skin; and as a result, the temperature of the body remains at a fixed point.

Effects of Lowering the Temperature. The body is warmed by heat generated within itself. To lower its temperature, it is only necessary to abstract the heat faster than it can be produced. The first effect of this is pain in the more exposed parts. The face and extremities "ache with the cold." This soon passes away and the skin becomes quite insensible. The testimony of individuals who have been rescued from freezing, even after they were insensible from the cold, is that a disposition to sleep overtakes them soon after the pain has left the skin; the muscles become inactive; breathing is slow and difficult; and the whole nervous system becomes sluggish. Finally, the desire to sleep becomes irresistible, and in a short time death ensues. When rescued from a freezing condition, it is found that respiration is hardly perceptible, the heart's action extremely weak, and all the functions of the body nearly suspended.

The above condition is very similar to a normal physiological process among the so-called hibernating animals. These animals go to sleep at the approach of winter, and do not waken until the coming of spring. When found, buried in their nests, or deep underground, they are quite insensible and immovable. Their respiration is hardly discernible, and their bodily temperature is much reduced. The oxidation of tissue is very slight, and the animal appears to live by using up its own flesh, — entering upon sleep well supplied with fatty tissue and awakening in the spring very poor.

Effects of Raising the Temperature. When the animal heat is raised a number of degrees, as in fevers, the effects are quite the reverse of those produced by cold. In fevers, the pulse and the respiration are increased in frequency, and instead of a feeling of comfort and sleep, there is often much distress and wakefulness. Increased temperature appears to hasten the normal changes taking place in the tissues; oxidation is more rapid; the tissues are more quickly exhausted, and the vitality is lowered.

Winter and Summer. The moderate cold of winter imparts a feeling of vigor and stimulates the whole system. The cool air excites a desire to run, and to exercise the whole body; this activity adds to the amount of heat necessary to resist the cold. Without exercise, the internal heat must be preserved by additional clothing, or the depressing effects of cold will be experienced. Cold weather brings a good appetite; the extra amount of

food is so much more fuel, contributing to the mainte-
nance of the animal heat; thus we learn that a healthy
body always demands more food during the winter than
during the summer. Muscular activity, extra clothing,
and more food enable the body to resist the cold and
still maintain its average temperature. During the heat
of summer less food is required and more liquids are
used, the perspiration is increased, and the clothing is
lighter, all of which tend to diminish the supply of heat
and increase the means for its escape.

The Effects of Alcohol on the Temperature. There is a
general belief among a large class of people that alcohol
warms the body; so they naturally conclude that it is
one of the best remedies to take before beginning a long
journey in the cold. If alcohol is able to raise the temperature of the body, it might often prove a most desirable agent. But has it this power? To fully answer
the question it is necessary to understand both the
primary and secondary effects of this poison. Understanding these, the reader will be able to draw his own
conclusions.

One of the first effects of alcohol is to paralyze the
nerves which terminate in the walls of the small blood
vessels, and which control their size. As a result, the
vessels distend and the whole system of capillaries becomes filled with an extra supply of blood. The minute
blood vessels of the skin become distended with this extra supply. Now it must be remembered that the blood
from the interior of the body is much warmer than that
near the surface; therefore when an extra amount of
warm blood flows to the surface, it makes the skin feel
much warmer. As the outside of the body feels warmer,

a person is led to believe that the warmth extends throughout the whole system. This is physiological, and just what we should expect. But we know that the blood is cooled in passing through the skin, therefore, if more than the usual amount of blood be sent to the surface, then an unusual amount of heat will be given off. As a result of this extra loss of heat, the blood returns to the interior of the body much lowered in temperature, and cooling all the tissues with which it comes in contact.

The period during which the skin feels warmer is very brief, while the period of cold following it is of much longer duration. During the second period there is a rapid fall in the temperature. This is physiological also, and just what should be expected. In some of the lower animals the decline may be very great, reaching as many as five degrees in birds, and three degrees in dogs. In man the decline is often as great as two degrees, in some cases even more. This period of decrease may pass off in a few hours, if the amount of alcohol taken be small; but if the poisoning be sufficient to cause prolonged sleep, several days may be required to bring the temperature up again to its normal degree.

From these facts it appears that the brief period of apparent increase of heat is followed by a longer one during which the temperature of the body is actually reduced. In this condition the body is less able to resist the cold, and so becomes especially liable to influenza, bronchitis, pneumonia, and allied affections; while recovery from their attacks will be more tedious and doubtful. In other words the resisting power of the body against all invading diseases is greatly weakened by the use of alcoholic beverages.

There is an abundance of testimony from men who have been exposed to extreme cold, to show that alcohol increases the suffering and danger of such exposure. Explorers in Arctic regions and travelers in cold countries are perfectly agreed on this question. Their testimony is all to the effect that the use of alcohol in cold countries is extremely hazardous. The same principle applies to exposure to unusual cold in more temperate climates, and shows the fallacy of drinking alcoholic liquors to " warm one up " or to aid in keeping warm on a cold day. They only make exposure to cold more dangerous. The men who never use alcohol bear such exposures much better and do their work more easily than those who take it.

QUESTIONS.

1. Give some of the sources of animal heat.
2. Where is found the warmest blood in the body?
3. Which are the cold-blooded animals? Name some.
4. Which are the warm-blooded animals?
5. In what parts of the body is the temperature the highest?
6. How does the blood tend to equalize the temperature?
7. What is the temperature of the healthy adult body?
8. How much heat does the body produce in one hour? In thirty-six hours?
9. What two tissues principally regulate the temperature?
10. Explain the part performed by the lungs
11. How does the skin regulate the temperature?
12 Give some of the effects of lowering the temperature.
13. This is similar to what normal process?
14. Describe the effects from raising the temperature?
15. What is one of the first effects of alcohol on the blood vessels?
16. What is the result of this first effect?
17. Does alcohol aid in resisting cold?

CHAPTER XXII.

THE EFFECTS OF TOBACCO.

WE believe that the desire among boys to be men and to do manly things is so great that if they knew the dwarfing and stunting effects of tobacco, they would never indulge in its use. For this purpose, therefore, we are anxious that the exact truth be clearly stated.

Our objections to the smoking of cigarettes, or tobacco in any form, are as follows : —

1. **It Lessens the Natural Appetite for Food and Injures Digestion.** We have already discussed the growth of cells, and the important part they play in all the processes of life. These rapidly changing cells must be fed ; but if the appetite be poor and digestion bad, then surely cellular action must be greatly affected. Such proves to be the case, for the early use of tobacco often checks the growth of the body, so that it never reaches the height and full development it would have reached, had it not been so poisoned.

2. **It Seriously Affects the Nervous System.** This is shown by its effects on the heart. The unsteady and rapid beat leads to dizziness and rush of blood to the head. The sleep is disturbed with distressing dreams, and the morning finds the body unrefreshed. The brain is checked in its development, and is unable at all times to do its best work. The testimony of many eminent

medical men, together with the statements of public
men of wide observation, show unquestionably that
Tobacco impairs the Mental Powers. Could a more
serious charge be brought against it? The evidence is
strong and unanswerable. Its moderate use by the
young weakens the ability to think, while its immoderate
use may eventually destroy the mind.

3. **It Leads to the Opium Habit.** It is a startling fact
that some of the brands of cigarettes have opium mixed
with them. The amount is small, — the tobacco may be
only moistened with some weak juice of the poppy; but
the active principle of the opium is there, and it will
surely perform its poisonous work. It is to be feared
that if our young men continue the use of cigarettes
we shall soon see, as a legitimate result, a large number
of adults addicted to the opium habit.

4. **It Creates a Craving for Strong Drink.** This is another legitimate result of smoking. It is a result which
could have been foretold simply by physiological reasoning. That smoking leads to drinking is no longer a
theory, for the wisest observers have testified that such
is the case. There are both primary and secondary
reasons why this result follows. It follows as the primary effect of tobacco on the mouth, throat, and stomach;
the mouth and throat are dry, and there is a peculiar
sinking sensation at the stomach; these often tempt the
smoker to drink. It is also to be noted that the
majority of those who use alcoholic drinks also use
tobacco; therefore when a person accepts an invitation
to smoke he is more likely to be thrown into company
with those who drink.

The secondary reason is that as tobacco weakens

the mental faculties and lowers the moral tone, so it
makes the temptation to drink more difficult to resist.
Smokers yield more readily to an invitation to drink.
They accept the " treat " to a cigar or a cigarette ; and
later the acceptance of a " treat" to strong drink becomes much easier. Tobacco manufacturers have lately
declared that since the more stringent temperance laws
of Canada have been enforced, there has been a vast re-
duction in the amount of tobacco consumed.

5. **Cigarette Papers often contain Arsenic.** The effects
of such a drug on a healthy organism cannot be otherwise than injurious.

6. **It is a Filthy and Offensive Habit.** The laws of
ordinary politeness are violated daily by those who use
tobacco in any form. A noted French lady, when once
asked if it would be offensive to her to have gentlemen
smoke in her presence, curtly replied, " No *gentleman*
ever smokes in my presence." The man who smokes
becomes completely saturated with the strong odor.
His clothing, his living room, and even his breath are
charged with it. It is astonishing to see how the
smoke of tobacco will nearly strangle the young lady as
she sits beside her friend in the carriage, or as she walks
by his side in the street, and yet she quietly submits to
the impolite treatment. Travel where you will, on the
road, in the railway car, on the sidewalk, in the street
cars, in halls, everywhere, you can see the traces of this
filthy and offensive habit.

7. **It is Expensive.** The purchasing of tobacco requires
money which might be otherwise spent for personal comforts and pleasures, or for public and benevolent enterprises. The trouble is that the young man does not

realize what it is to spend five or ten cents each day. Such small sums, placed at compound interest, grow into large figures by the time old age is reached. It would be far better to invest these savings in books, newspapers, lectures, concerts, or traveling.

8. **It is Unlawful.** The statesmen of our country are so fully impressed with the fact that the coming generation is likely to be weak in body and mind unless the use of tobacco by the young be checked, that at this time in twenty-nine States there are stringent laws forbidding the furnishing of cigarettes or tobacco in any form to minors under certain ages. There are also at this time in thirty-five States and in all the territories, laws requiring instruction to be given to all pupils in the public schools on the nature and effects of tobacco, as well as of alcoholic drinks and other narcotics.

In New York and Connecticut, it is not only unlawful to furnish tobacco to persons under sixteen years of age, but it is also unlawful for such persons to smoke or use tobacco in any public place. In Massachusetts and Indiana, a person can be fined for advising or persuading any one under sixteen years of age to smoke or chew tobacco. In the District of Columbia, 257 physicians, 524 officers and teachers of the public schools, all the trustees of the public schools, and 86 pastors of churches, petitioned the 51st Congress for the passage of a bill prohibiting the selling, giving, or furnishing tobacco in any form to persons under sixteen years of age. Individuals can now walk the streets of the Capital of the nation and not see a single newsboy or other young lad with a cigarette in his mouth. The simple fact that the representative men of a whole nation, in Congress assembled, should pass such a bill, shows in itself that a grave danger was discerned, and that the time had come for the remedy to be applied.

9. **It Lowers Scholarship.** The minister of Public Instruction of Paris has forbidden the use of tobacco by students of the public schools. Tobacco is prohibited in the military and naval schools of the United States Government.

Notice the following testimonies, which could be multiplied many times : —

" In our thirty years experience in teaching more than fifty thousand young people we have found the effects of this narcotic to be premature age, shattered nerves, mental weakness, stunted growth, and general physical and moral degeneracy ; and therefore *we now decline to receive into our institution any who use this noxious weed.*"

Such is the statement of Henry C. and Sara A. Spencer, principal and vice-principal of the Spencerian Business College.

Dr. Willard Parker says, "Tobacco is ruinous in our schools and colleges, dwarfing body and mind."

The President of the Baltimore Academy of Medicine says, " The effects of tobacco on school-boys are so marked as not to be open for discussion."

In an examination for admission to the Free College of New York, seventy-one per cent of the girls who applied were passed, but only forty-eight per cent of the boys. The report attributes the smaller per cent of the latter to the stupefying effects of tobacco.

A graduating class in Yale College was divided into four parts according to scholarship, — the best scholars in the first division, and the poorest in the fourth. In the first division twenty-five per cent used tobacco; in the second, forty-eight per cent; in the third, seventy per cent; and in the fourth, eighty-five per cent.

Dr. J. W. Seaver, of Yale College, has made some interesting observations on this subject. He says no young man can use tobacco without injuring himself seriously. Of the junior students who received the highest appointments, ninety-five per cent did not use tobacco; of those who received the next highest appointments, eighty-seven and one half per cent did not use it; while of all who received appointments of some kind, eighty-four and three-tenths per cent did not use tobacco. The per cent of non-users among those who received no honors whatever was much less. In other words, the highest honors of a class are almost exclusively carried off by those who do not use tobacco; while those who remain lowest in scholarship are almost invariably addicted to the habit. This is in accord with the testimony of a large number of teachers of wide experience, who positively assert that the pupils who use tobacco are not found at or near the head of the class. Tobacco and scholarship are no friends. They cannot go together.

The President of Union College says, " The use of this poisonous narcotic, next to intoxicating liquor, is more destructive to the health˜of our youth than any other agent."

The Emperor Louis Napoleon, in 1862, issued an edict forbidding the use of tobacco in all the national institutions, because he was satisfied from investigation in the schools and colleges that the average standing in both scholarship and character was lower among those who used tobacco than among those who did not.

10. **It Lowers the Moral Tone.** This is most marked in the deceit practised by boys in their efforts to conceal the fact of their using tobacco from their parents. They will positively deny its use, and resort to all manner of deception to account for the unmistakable odor. Boys who would not be guilty of telling a falsehood on other matters soon find it easy to lie about this habit. They hide their cigarettes ; they go away from home to smoke them ; and in all their manners show that they are in a bad business. What result can follow such falsifying and deception other than a lowering of the moral tone ?

The ten charges thus brought against tobacco are of such a character that there is but one conclusion to the matter. It is this : The use of tobacco greatly endangers and impedes health, happiness, and prosperity.

QUESTIONS.

1. How does tobacco affect the appetite and digestion ?
2. How does it affect the heart ? The mental powers?
3. Is there danger it may lead to the opium habit? Why ?
4. Give a primary reason why smoking leads to drinking.
5. Give a secondary reason why this is true.
6. What poison is sometimes found in cigarette papers?
7. Are politeness and neatness aided by the tobacco habit ?
8. Is it an expensive habit?
9. What is said about its being unlawful?
10. Is tobacco an aid to scholarship ?
11. What proof have we that it affects the morals?

CHAPTER XXIII.

WHAT TOBACCO COSTS: IN BODY; IN MIND; IN MORALS; AND IN MONEY.

What it Costs the Body. It is not natural for man to smoke. There is not an animal in a state of nature that uses tobacco in any form, unless exceptions are made in favor of the tobacco-worm, living on the green leaves of the plant, and a species of wild goat found in Africa. That it is not natural for man is evident from the shock to the system which follows its first use, resulting in nausea, vomiting, and great depression. Upon experiencing these symptoms the first Napoleon exclaimed, " Oh, the swine! my stomach turns!" Man is the only animal using this vegetable. How different, in this respect, is tobacco from other vegetables. If we eat half of a potato, apple, carrot, beet, or any other vegetable, raw or cooked, and then throw away the other half, how eager are many of the lower animals for it! If the horse and dog do not care for it, then hosts of ants and flies claim it as their own. But not so with the half-smoked cigar, carelessly thrown to the ground. The horse, dog, ants, flies, even all forms of animal life below man shun it, and there it remains where thrown, until the street-sweeper or the storms have removed it. Man alone appears to care for it. Tobacco is a disease-producing instead of a disease-curing plant. In the preceding chapters attention has been called to the ill effect of this drug on the growing body. It has been shown that its effects on the appetite and digestion result in checking the growth and in producing a poorly developed organism. Chronic sore throat, with its accompanying deafness, are often a part of the smoker's complaints. The palpitation of the heart, the trembling of the hands, and the excessive nervousness all testify that tobacco is used at an immense waste of vital energy.

But let us listen to the teachings of some of our eminent writers on this subject. Below are given a few extracts, a number of which were taken from the writings of " Meta Lander," — Mrs. Margaret Woods Lawrence, of Baltimore, — entitled " The Tobacco Problem." The gifted author has collected a vast amount of material bearing on this important subject. The quotations represent the convictions of men of high standing and acknowledged ability.

Dr. Ferguson : " I believe that no one who smokes tobacco before the bodily powers are developed ever makes a strong vigorous man."

Many leading physicians of Philadelphia : " Cigarette smoking is one of the vilest and most destructive evils that ever befell the youth of any country ; its direct tendency is to a deterioration of the race."

Prof. H. H. Seerley : " Boys that begin the habit at an early age are stunted physically, and never arrive at normal bodily development."

Horace Greeley : " Show me a drunkard that does not use tobacco, and I will show you a white blackbird."

Medical Director U. S. Navy: "The future health and usefulness of the lads in our naval schools require the absolute interdiction of tobacco in any form."

Dr. Copland, F. R. S.: "Tobacco creates a thirst, to remove which alcoholic stimulants are often resorted to."

Dr. Cowan: "The exceptions are very rare when a user of tobacco in any of its forms is not ultimately led to use alcoholic liquors."

Gladstone "detests smoking." Haeckel: "I have never smoked." Ruskin "abhors the practice of smoking."

Charles Reade: "I have seen many people the worse for it, and never saw anybody perceptibly the better for it."

Dr. Willard Parker: "Tobacco is doing more harm in the world than rum. It is destroying our race."

Medical Examiner U. S. Navy: "One out of every one hundred applicants for enlistment is rejected because of irritable heart, arising from tobacco poisoning."

Dr. Bowditch: "A man with a tobacco heart is as badly off as a drunkard."

The famous marksman, Carver: "I have never tasted intoxicating drinks, nor do I use tobacco in any form."

The winner of the international boat race, Hanlan: "The best physical performances can only be secured through absolute abstinence from alcohol and tobacco."

Supt. of N. Y. Insane Asylum: "Tobacco has done more to cause insanity than spirituous liquors."

Supt. Insane Asylum at Worcester, Mass.: "That tobacco produces insanity, I am fully convinced."

In four insane asylums there were 294 cases of "insanity from the use of alcohol." Of these 294, it was ascertained that 206 were led to intemperance by the use of tobacco.

The German Government has ordered the police to forbid boys under sixteen years of age smoking in the streets, because of its evil effects on mind and body.

These facts are sufficient to warrant us in stating that tobacco costs the body its best development and its perfect health.

What it Costs the Mind. Reference has already been made, in the preceding chapter, to the effects of tobacco on the general scholarship. Some further testimony on this point may serve to more forcibly present the truth.

In Paris " it is shown that smokers have proved themselves, in the various competitive examinations, far inferior to others."

Dr. Constan, of Paris, concludes a long article as follows : "The influence of tobacco clogs all the intellectual faculties, and especially the memory ; and the injury is greater in proportion to the youth of the individual."

Ex-Senator Doolittle: "I believe that the mental force, the power of labor and endurance of our profession, is decreased at least twenty-five per cent by the use of tobacco."

Professor Lizars, of Edinburgh : "It is painful to contemplate how many promising youths must be enfeebled in their minds and bodies, before they arrive at manhood, by the use of tobacco."

With these facts before us, it is safe to assert that tobacco costs the mind its highest development and its most brilliant achievements.

What it Costs the Morals. The use of tobacco is an illustration of the most supreme selfishness. One person only enjoys it, and even that enjoyment is usually at the expense of those about him. The breath, the clothing, the air around the smoker, all are saturated with the tobacco-poison. The fact that generous and kind men will continue in such a purely selfish practice is simply an illustration of the truth that the drug has so affected their sense of propriety as to render them unaware of their true position. If men must smoke they should confine their operations to their own private quarters, where as few as possible must suffer, rather than be continually puffing their cigars in the faces of enduring women and inoffensive children.

Upon entering depots, halls, and other public places, we are often greeted with the sign, " No Smoking." It seems to be absolutely necessary to give stern commands to men not to smoke in these places where women and children gather. Yet why do we not see commands like these? " Do not crowd the children." " Do not thrust the sick out into the cold." In other words, were it not for the effects of tobacco on the moral sense, blunting its keener qualities, there would be no need of asking men not to smoke in any place where the rights of others to the free, pure air are universal.

We positively declare that a man has no more right to put tobacco smoke into the air which we are about to take into our system, than he has to put some disagreeable drug into the water which we are just about to drink. Pure air is one of our inherited gifts, and he who wilfully poisons it, or makes it disagreeable for his fellow beings, is guilty of no small breach of common civility.

On the morals of the youth, tobacco has a most marked effect. It is probable that the indifference of the adult smoker is largely due to the early use of the cigar or cigarette, as during the early years of life all the faculties are so much more easily affected. Deceit seems to be a boon companion of the boy and his cigarette. To have the morals of our young men reach their highest attainments, the cigarette should be banished from our midst. What do some of our writers and public men say on this subject?

Jackson Jarves says, " Whatever the benefit or harm the use of tobacco may do the consumer's body, its common tendency is to render the mind indifferent to the well-being of his neighbors."

Dr. Parker: " Tobacco demoralizes. It makes a man careless about his hair; he lets his nails go unclean; his clothes are soiled, and he is generally untidy."

Professor Stuart, of Andover: " It creates a nervous irritability, and thus operates on the temper and moral character of men."

Professor Mead, of Oberlin: " The tobacco habit tends to deaden the sense of honor; and none are more likely to practise deception unscrupulously than those who use the weed."

The President of Wisconsin State University : " There are few spectacles giving a more disgraceful impression of our civilization than that of a lad sporting a cigar or cigarette, in imitation of the bad habits of those older in years than himself."

Superintendent of the Reform School, at Westboro', Mass. : " All boys sent here have been the users of tobacco."

Chancellor Sims, of Syracuse University : " The tobacco-habit is deteriorating to the one indulging in it."

Dr. Harris: "There is not another article of luxury that so secretly and yet so surely saps all the foundations of manliness and virtue."

From these statements we can but draw the conclusion that he who indulges in tobacco runs the risk of having the highest and best attributes of his nature brought under its benumbing power."

What it Costs in Money. The financial view of the tobacco problem merits careful attention, not only because the money expended for the cigars or cigarettes might be better used in the purchase of books, but also because habits of thrift and economy formed in the young make no small part of a man's future success. Such habits do more than make a saving of money; they teach that time itself is precious, and that rapidly passing hours can never be recalled. The normal physiological man is economical of his money, his time, his mental powers, and of all his endowment. He is temperate in all things. He should have immense reserve forces which he draws upon only at times of great peril and need.

Later in school life our readers may find themselves hard at work over some of the great questions of political economy. In fact, even now they are aware there is much unrest in the land over the question of capital and labor. A small number of people are becoming richer and richer, and controlling great interests and thousands of men, while a large number of people work very hard, year after year, and yet do not enjoy the ordinary comforts of life, to say nothing of owning a home of their own. Often it is true that sickness and misfortune keep the most economical persons from laying aside any part of their hard-earned wages. But we desire to show, not only that the use of tobacco injures the consumer, but also that the money thus wasted might bring many additional comforts to him and to his friends, thus contributing much toward the general prosperity of the country. True patriotism, therefore, demands that boys should not use tobacco. A few lines will show how expensive is the tobacco-habit, and how important a part it plays in preventing many of our hard-working people from enjoying more of the comforts and pleasures of life.

The secret of the trouble lies in the fact that we fail to realize how soon cents makes dimes, and dimes make dollars. Remember that, even at the rate of six per cent compound interest, money will double itself in twelve years, while if the rate be ten per cent, money will double in seven years. This doubling of money, without any apparent effort on our part, is one of the surprising things of finance. Let us illustrate this principle.

Suppose a boy should begin the use of tobacco at ten years of age, and suppose he purchases only one cent's worth of tobacco each day, until he is forty years of age; now had he received seven per cent compound interest for the money thus expended, how much would his tobacco bill represent? $350. Neither boys nor men, though, stop at a cent's worth of tobacco a day. Suppose he used only five cents a day for his tobacco during all these years, still this, at the above rate of interest, would amount to over $1700; surely a good present to receive on your fortieth birthday! But suppose the young man is now twenty-five years of age and just starting out in business for himself. He buys five or six good cigars each day and they cost him fifty cents. This amount regularly placed at seven per cent compound interest for thirty years would make his tobacco bill represent over $17,000. A man spends a dollar a day for tobacco. How much would his tobacco bill represent in only ten years, at seven per cent compound interest? over $5,000. In twenty years? over $16,000.

We have a very interesting letter on this subject, from Luther Prescott Hubbard of New York. Mr. Hubbard has been Secretary of the New England Society for nearly forty years, and for nearly sixty years has been the financial agent of the American Seamen's Friend Society. His connection with literary and financial enterprises entitles his statements to great weight. Mr. Hubbard began to chew tobacco at the age of twelve, and a few years later commenced smoking. He says his tobacco cost him, on the average, thirty-seven and a half cents each day. We will let him tell the rest in his own language which he kindly allows us to use: —

"I now deposited the money I had been so long squandering for tobacco in the Seamen's Bank for Savings. I will tell the boys what I did with it, that they may see how unwise and inexpedient it is to commence the expensive, demoralizing habit of smoking or chewing tobacco.

"We had long lived in the city, but the annual visits of the children to their grandfather's made them long for a home among the green fields. I found a very pleasant place for sale. There were over two acres of land, with abundant shade and fruit trees, a good garden, a fine view of Long Island Sound, — near the academy, churches and schools, and a convenient distance from New York. The cigar money was drawn upon to purchase the place, and it is mine.

" My smoking was moderate compared with that of many, costing me only thirty-seven and a half cents a day, equal to $136.50 per annum, which, at seven per cent interest for *fifty-nine years*, amounts to the small fortune of $103,626.32. This has afforded means for the education of my children, with an appropriate allowance for benevolent objects.

" Great as this saving has been, it is not to be compared with improved health, a clear head and steady hand, at the age of over *eighty-three years*, and entire freedom from desire for tobacco in any form."

There are other practical questions that should make every intelligent schoolboy appreciate the money part of this evil. If a boy saves fifteen cents each day while going through his four years' course in the high school, how much money will he have? Suppose this money had been invested in books, and suppose that he had purchased twenty-five volumes of the most recent works at an average price of two dollars each, and fifty volumes at one dollar each ; in addition to these how many volumes of the standard works of such writers as Scott and Dickens, Gibbons and Macaulay could he purchase at thirty-five cents each? " A good sized library," you will say, when you figure the result. Too choice a collection to be burned up at the rate of fifteen cents a day.

Only one verdict can be given on this subject. It is that, after the most careful examination of the whole tobacco question, *not a single fact can be brought forward in its favor*.

CHAPTER XXIV.

OPIUM. — CHLORAL.

The Opium Habit. In this country the opium habit is usually formed as a result of taking the drug for the relief of pain. If the pain returns, another dose is taken. Sometimes the disease is of such a character that the pain continues for many weeks, necessitating the continued use of the opium. After a time, however, the original disease may entirely disappear, yet the sufferer finds he must continue the opium, or endure great agony of body and mind. Knowing how easily the opium habit is contracted, the careful physician does not prescribe it for those conditions where its use is likely to be necessary for some weeks. He substitutes less powerful narcotics, and allows his patients to suffer some pain, rather than run the fearful risk of producing such a terrible habit.

Effects of Opium. Opium is a powerful drug. Even a single dose prescribed by a physician for the relief of some severe and sudden pain is often followed by undesirable effects. But when repeatedly taken for a long time, the body becomes thin, and the skin grows sallow and parchment-like. The opium eater suffers from loss of appetite and improper action of all the organs of the body. There is a marked lowering of the will-power, and loss of memory. The person will not hesitate to deceive and lie about the habit. He loses all sense of truth, and appears to forget there is any difference between right and wrong. When unable to obtain the drug, the victim suffers the most excruciating physical and mental torments.

Preparation of Opium, or Patent Medicines. It is a well-known fact that nearly all physicians are opposed to so-called patent medicines. The chief reason for this is that, as they are ignorant of the composition of the medicine they are unable to pass judgment upon its merits and demerits. When a person is ill the physician prescribes those drugs which will have certain desired effects in that particular case. It may be an acute cold, and the very next patient he visits may also be suffering from the same affection; yet entirely different drugs may be prescribed in the second case. But a patent medicine makes it necessary that persons of different constitutions, living in various climates, at all seasons of the year, and in varying degrees of health, or in a word, that all classes and conditions of people everywhere must take the one unvarying mixture. No one man can intelligently prescribe a single mixture which will be applicable to thousands of cases; it is as impossible as it is unscientific.

A large number of the popular " soothing syrups " of the present day contain opium. These syrups are freely given to children to produce sleep, or to correct some disturbance of the digestive organs. They are directly responsible for many deaths. Nothing of the kind should ever be given. Let all such cases be referred to a physician. Is it not more probable that your own local physician should know better what is needed in case you are ill than a man, uneducated in medicine, living thousands of miles away ? The " tonics," " bitters " and " soothing syrups " of the market are too likely to contain either alcohol or opium to make them safe remedies.

Chloral. A few years ago this drug was extensively prescribed by medical men because it was capable of producing sleep, relieving pain, and appeared to be free from danger. After some experience with it, the discovery was made that it was far from being free from danger. A number of deaths were reported from its use, and many persons became addicted to the chloral habit. Of late it is not used so extensively, yet there is danger that it may be taken repeatedly as a substitute for opium. The habit is easily acquired and as difficult to break as that of taking opium.

After a short use of chloral the tongue becomes coated, the digestion impaired, and the stomach unable to retain the food. There is oppression and pain in the stomach, associated with nausea and vomiting. The nervous system soon breaks down under the constant use of the drug; the person becomes irritable, the muscles tremulous, and the heart irregular in its beat. It soon becomes impossible to procure sleep without large doses of the drug; while it not infrequently occurs that too large doses are taken, with fatal results.

Tea and Coffee. The active, growing body finds in these substances nothing which it needs. Tea and coffee are especially bad for those who have nervous temperaments, and those who lead indolent lives. They interfere with digestion, and often produce biliousness and wakefulness. The student can accomplish more and better work without these drinks than with them.

CHAPTER XXV.

THE ANATOMY OF THE NERVOUS SYSTEM.

Two Systems. The nervous system of man and all vertebrate animals may be divided into two secondary systems; each having its own particular set of nerves and nerve centers. These are called the cerebro-spinal system, and the sympathetic, or ganglionic, system.

The Cerebro-Spinal System. The cerebro-spinal system is composed of the brain, the spinal cord, and the nerves which originate from them. The brain and the spinal cord are the great nerve centers of the body. They are connected with the nerves of the special senses, and with the nerves of common sensation; they convey to the mind the sensations of taste, touch, sight, smell, and hearing, as well as the sensations of pain, hunger, thirst, etc. The mind, in turn, is capable of expressing itself through them. Through the cerebro-spinal system the commands of the mind are conveyed to various parts of the body; thus, we " will " to move a muscle; instantly a force is sent along the nerves of this system to the proper muscle, and it promptly obeys.

The Sympathetic System. The sympathetic, or ganglionic, system consists of a number of ganglia and nerve fibers. The ganglia are collections of nerve cells. They may be very small, composed of only a few cells, and visible only with the microscope; or there may be such a collection of nerve cells that the ganglia have considerable size. The sympathetic system consists of two rows of these ganglia, one row on either side of the spinal column. It consist also of nerves which proceed from the ganglia to the organs in the thoracic, the abdominal, and the pelvic cavities. The sympathetic nerves do not go to the skin, neither are they connected with the special senses, nor are they under the control of the will. This system presides over the involuntary processes of the body; as the circulation, the digestion, the respiration, the absorption, the nutrition, and the involuntary muscles. It also controls the secretions of glands, and has much to do with the amount of blood distributed to the various organs and tissues.

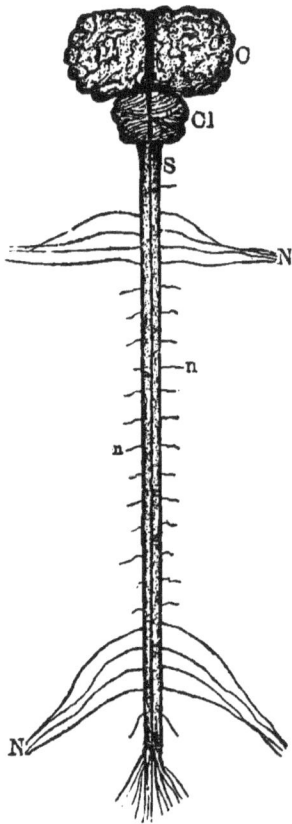

FIG. 89. Diagram of the cerebro-spinal system : c, cerebrum ; cl, cerebellum ; s, spinal column ; N, nerves for the upper and lower limbs ; n, nerves going to the muscles and the skin.

The ganglia are connected with each other by small nerve fibers, while other minute fibers connect the ganglia with the cerebro-spinal system. Thus one organ is made to act in "sympathy" with another, so that if one part suffers in any way another suffers with it.

Nerve Tissue. There are two kinds of nerve tissue; one is composed of nerve fibers, and the other of nerve cells. The presence of numbers of the nerve cells gives a gray color to the tissue, while the nerve fibers appear white. Hence an accumulation of nerve cells is called the gray substance, and a collection of nerve fibers is called the white substance. The nerve cells represent the centers of activity from which the orders are issued. The nerve fibers are simply the conductors, conveying the messages from place to place. The gray substance therefore, represents the seat, or origin, of the mysterious forces of the nervous system; while the white substance represents only so many fibers for the transmission of the forces.

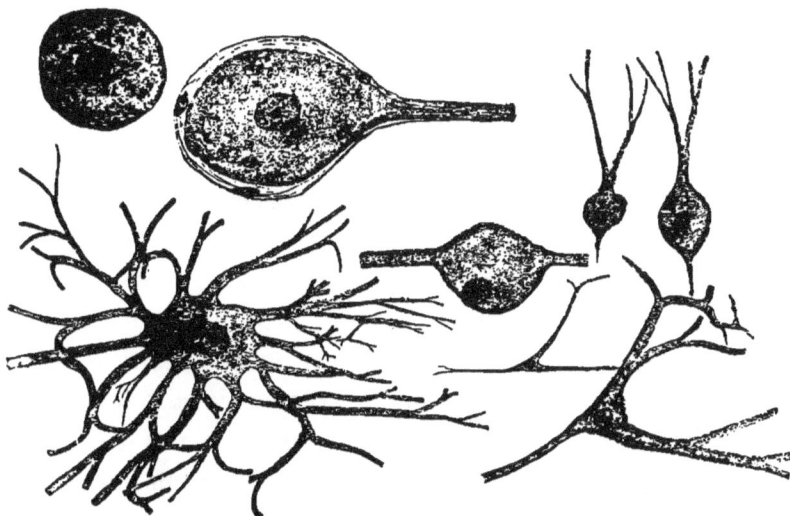

FIG. 90. Various forms of nerve cells, highly magnified.

The Nerve Cells. A collection of nerve cells always makes a nerve center. The greatest collection of cells is found in the gray matter of the brain, the next in the spinal cord, and the next in the ganglia of the sympathetic system. Nerve cells generate nerve force. They also receive it from other cells and give it out again. They are therefore generators and transmitters of nerve force.

The cells vary exceedingly in size and shape. Some are of a circular form, as at the upper left corner of Fig. 90. Others have a long process extending from them, as shown in the cell to the right of the last. This one process represents the beginning of a nerve fiber, so that nerve force originating in the cell can be conveyed from it through this fiber to some distant organ or tissue. Other cells have many processes, one of which conveys nerve force to the distant parts.

The Nerve Fibers. The nerve fibers convey the nerve force from one part of the body to another. If an impression is made on the ends of some of the fibers they will convey it to the nerve cells at their origin : thus if the skin be touched, an impression is made on the ends of nerve fibers which is instantly conveyed to the spinal cord, and then to the brain, where a sensation is produced corresponding to that on the skin. No change occurs in the nerve fiber during the passage of this nerve force, so far as we know, any more than a change is produced in a wire by sending over it a telegraphic message. We know that if the wire be cut the message will be interrupted ; so if a nerve fiber be cut in any part of its course all communication is at once shut off.

Two Kinds of Nerve Fibers. The nerve fibers of the cerebro-spinal system are peculiar in this respect ; they can convey a message only in one direction. A message may be sent in either direction over a telephone or over a telegraph wire ; two wires are not necessary ; but with the nerve fibers, two complete sets are required. One set conveys sensation from the outer parts toward the nerve centers ; while the other set conveys impressions from the nerve centers outward to the muscles. The first set is composed of nerve fibers called sensory fibers, because they convey sensations to the spinal cord and brain : the other set consists of fibers called motory fibers, because they convey the stimulus of motion from the brain and spinal cord to the muscles. So far as is known there is no difference in the structure of these fibers, and as a rule they are side by side throughout the body.

A nerve, as seen in the body, and as mentioned in the books, consists of a large number of these nerve fibers held together by a delicate connective tissue. Each nerve fiber extends the whole length of the nerve, from its beginning in the brain or spinal cord to its termination.

The Brain. The brain is well protected in the cranial cavity. It is surrounded by three distinct membranes, of which the middle one is capable of secreting a fluid. The membranes protect the brain from friction against the bony walls, while the watery secretion gives it some freedom of motion. The brain is well supplied with blood vessels, — large arteries entering the base of the skull with the spinal cord.

The weight of the brain depends partly upon the size of the individual, and partly upon his intellectual capacity. The average weight of the brain of an adult male is a trifle over three pounds, 49½ or 49¾ ounces. The brain of Cuvier, the naturalist, weighed over 64 ounces, and the brain of Daniel Webster weighed 63½ ounces. The brains of idiots are very light, weighing from 27 ounces to as low as 8 ounces.

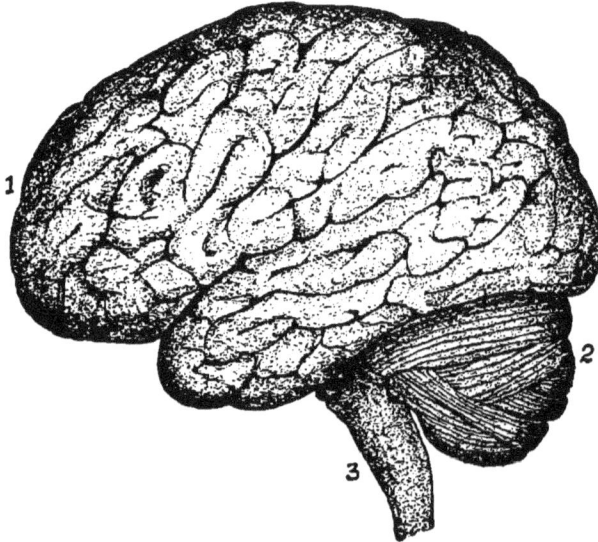

FIG. 91. Side view of the whole human brain : (1) cerebrum ; (2) cerebellum ; (3) medulla.

While it is true that many noted men had large brains, it is also true that some equally as noted had small brains. Yet it can be safely asserted that, as a general rule, the larger the brain the greater the capacity for intellectual power ; but to this statement there are some striking exceptions. It appears that there is something besides quantity necessary for the highest mental capabilities. This may be called quality, or some peculiarity of the brain material whereby it renders its possessor capable of great intellectual attainment.

The brain is divided into the cerebrum, the cerebellum, and the medulla oblongata.

The Cerebrum. The cerebrum is the brain proper. It

is the part above the ears, and is familiarly known as the great brain. It is believed that the cerebrum is the organ of the mind; that it is here we think, know, and reason. The cerebrum is divided into two parts by a

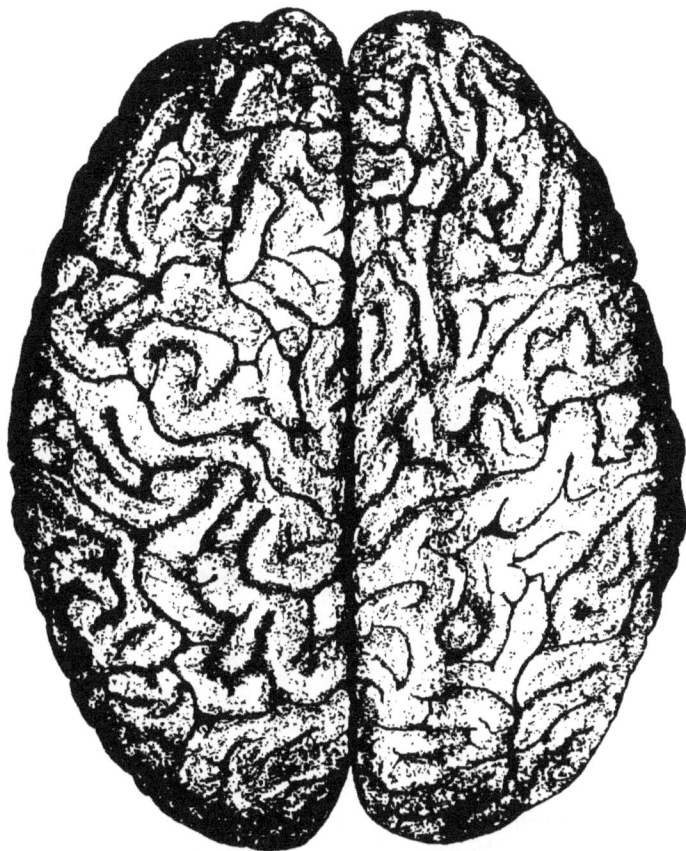

Fig. 92. The human brain, viewed from above: only the cerebrum is seen, with its deep fissure nearly dividing it.

natural fissure which passes from the front backward. At the bottom of the fissure the two parts are united by a band of nervous tissue, as shown at 3 in Fig. 93. From this it would at first appear that we have two brains corresponding to the right and left sides of the body; but doubtless the band of union between them not only connects the structure of the two but also in some way unites their functions. Figs. 91, 92, and 93 show that the surface of the brain is not smooth in

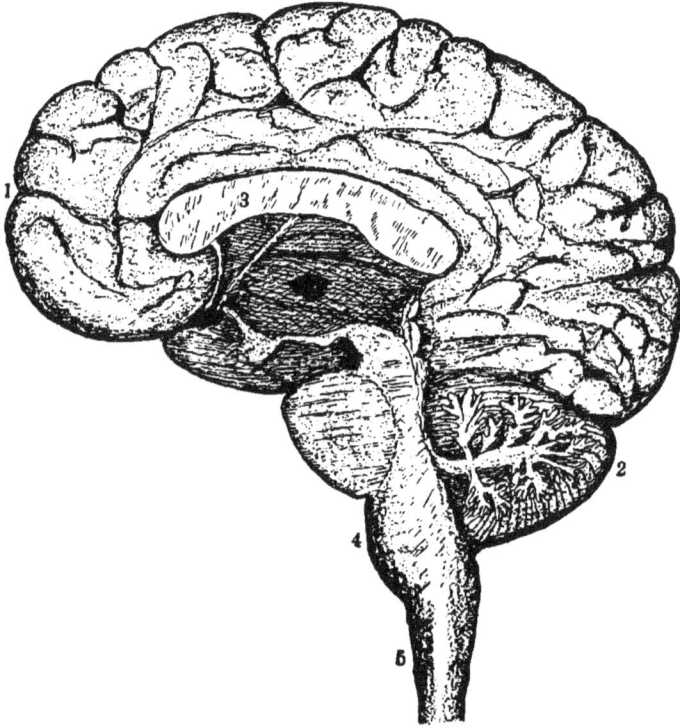

Fig. 93. One half of the brain, — the inner surface : (1) the cerebrum ; (2) the cerebellum ; (3) the band of tissue that unites the two sides of the brain ; (4) the medulla ; (5) the spinal cord.

man, but is thrown into a number of ridges or convolutions. The number of convolutions and the depth to which they reach vary in the different animals. In some, the surface is perfectly smooth, as in the pigeon and frog ; in others, the convolutions are shallow and few in number ; but in man, they are many and very deep. As a rule, the more intelligent the animal so much the more numerous and so much the deeper will be the convolutions of the cerebrum. The reason for this is readily understood when the structure of the brain is made clear.

Gray and White Matter of the Cerebrum. The gray matter is on the outside of the brain: the white matter is within, forming the center of the brain. The white matter is raised in slight folds on its surface to form the center of the convolutions; but the bulk of the convolutions is formed by the gray matter. It has been stated that the gray matter consists principally of nerve cells, and that these cells are the active agents in originating,

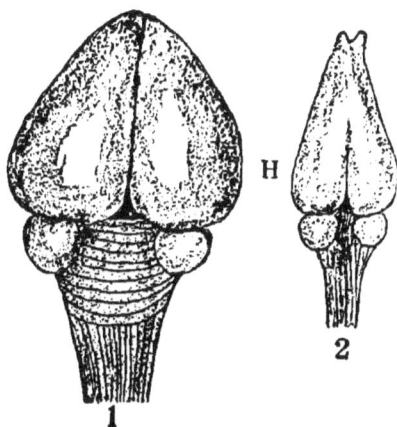

FIG. 94. (1) the brain of a pigeon; (2) the brain of a frog,— both viewed from above. There are no convolutions on the cerebrum, H.

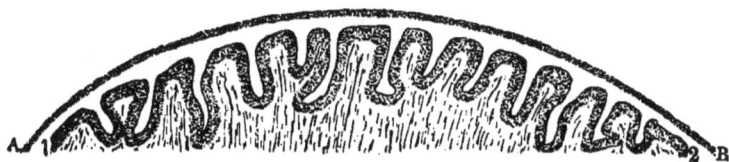

FIG. 95. A diagram illustrating that the convolutions of the brain give more surface for the gray matter.

receiving, and sending forth orders. The cells command and the fibers obey; the cells originate, and the fibers carry the messages. If the cells are specially concerned in originating and commanding, it is evident that a large amount of the gray matter is most desirable.

The convolutions provide for this extra amount of gray matter. This is made clear by the accompanying diagram. Suppose the surface of the brain be smooth, and covered with a layer of gray matter, then the line from A to B would represent the extent of the surface. But suppose the layer of gray matter be thrown into folds, or convolutions, then the amount of surface would be represented by the line 1 to 2. It is at once clear that the line 1 2 is much longer than the line A B. In other words, the convolutions greatly increase the amount of gray matter. It follows, therefore, that the deeper the convolutions and the greater their number, so much the more will the gray matter be in excess. This anatomical fact probably explains why some smaller brains are more intellectual than others which exceed them in size. In the former cases it is probable that the convolutions are deeper and more numerous than they are in the latter; thus actually giving more gray matter in the smaller brain.

The Cerebellum. This part of the brain is situated beneath the back part of the cerebrum, and is often called the lesser brain. It consists of gray and white matter arranged in the form of parallel ridges and furrows running over its surface, as represented in Fig. 91.

The Medulla Oblongata. The medulla, as it is generally called, is situated at the upper end of the spinal cord, between the cord and the brain. It represents an enlargement of the upper part of the spinal cord, as illustrated at 4, Fig. 93. It is well protected in the thick bones at the base of the skull. The functions of the medulla are so necessary to life that it must be regarded as the most vital portion of the entire body, yet it is only about one and one fourth inches in length.

The Spinal Cord. The spinal cord, as illustrated in Fig. 89, represents the elongated part of the cerebro-spinal system. It is about eighteen inches in length, one half an inch in thick-ness, and is nearly circular in shape. It is surrounded by three membranes which are continuations of those surrounding the brain, and is well protected in the spi-nal canal of the vertebral column. It begins at the medulla, and terminates at the lower end of the spinal column in a number of fine threads, as illustrated in Fig. 89. It is, like the brain, divided into halves by deep fissures. One fis-sure extends down the an-terior, and the other down the posterior median lines, nearly dividing the cord into two parts. Fig. 96 illustrates these fissures; one, the anterior, showing more clearly than the

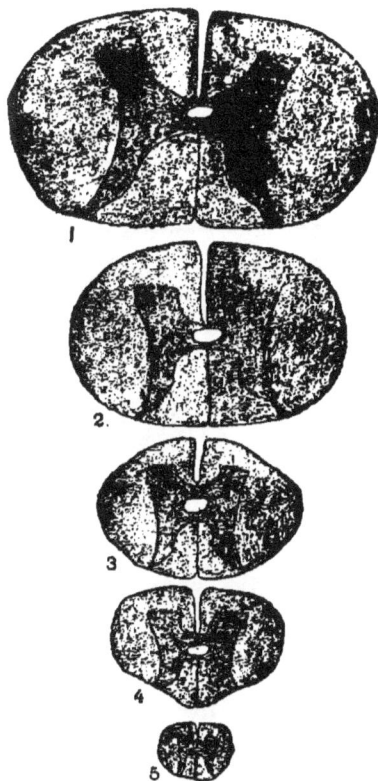

FIG. 96. Cross sections of the spinal cords of different animals, represented as twice the natural size: (1) horse; (2) ox; (3) man; (4) hog; (5) squirrel.

other. An open central canal is also seen. From the sides of the spinal cord there are given off thirty-one nerves, known as the spinal nerves.

White and Gray Matter of the Spinal Cord. The gray matter of the cord is in the center. It is so arranged that when the cord is cut transversely it slightly resembles the letter H. The darkly shaded portions in Fig. 96 illustrate this fact, and it is also shown in Fig.

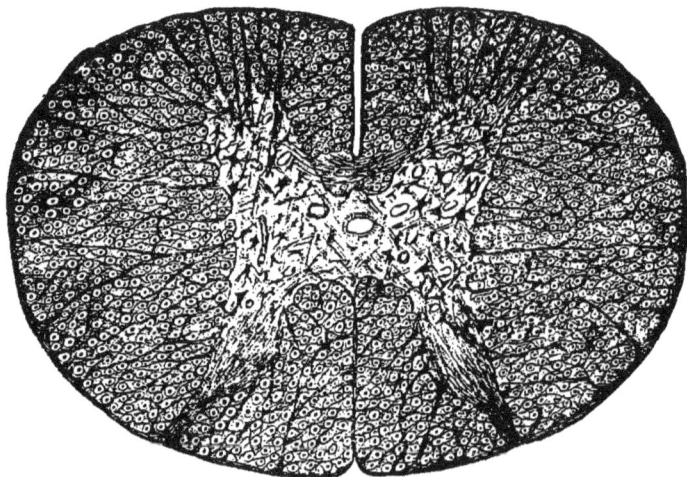

FIG. 97. A cross section of the spinal cord, magnified.

97. In the upper, or anterior points of the letter H, in Fig 97, are seen some irregular-shaped, darkly shaded bodies; these are the large nerve cells, a representation of which is given in Fig. 90.

Outside the central gray matter is the white matter, which is composed of fibers. The fibers extend up and down the cord, so that a transverse section of the cord, as seen in Fig. 97, shows the fibers to be circular and white, with a dot or dark spot in the center. This dark spot represents a cross section of that part of a nerve fiber which transmits the nerve force; the white substance around it is for insulation and protection. These fibers finally enter the brain and are distributed to all parts of it, being at last connected with the nerve cells.

The Spinal Nerves. Each one of the spinal nerves originates in the spinal cord by two roots. One root, the anterior, originates from the anterior part of the gray matter, as illustrated at 1, Fig. 98. The other originates from the posterior part, as shown at 2. Both these roots unite to form one nerve, at 3. On the posterior

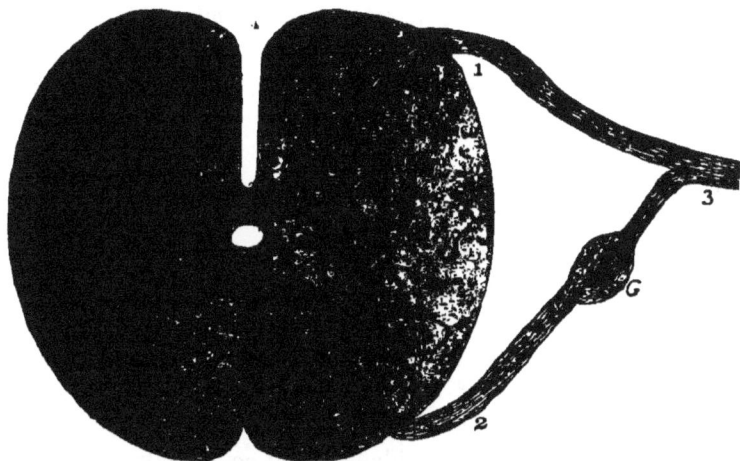

FIG. 98. A diagram illustrating the origin of the spinal nerves from the spinal cord · (3) is a spinal nerve; (1) and (2) are the roots, which originate from the gray matter of the cord; G, a collection of nerve cells; (1) the motory root, (2) the sensory root.

root is a ganglion, G, or a collection of nerve cells. The nerve fibers, at 3, continue together as a spinal nerve until they reach the distant parts of the body, when they separate again. Those fibers which originated from the anterior part, known as motory fibers, terminate in muscles; while those from the posterior part, known as sensory fibers, terminate in the skin.

QUESTIONS.

1. Name the two nervous systems.
2. What composes the cerebro-spinal system?
3. What composes the sympathetic system? What are ganglia?

4. What does the sympathetic system preside over and control?
5. Give the two kinds of nerve tissue.
6. How do they differ in color?
7. What do the nerve cells represent? The nerve fibers?
8. What makes a nerve center?
9. What do nerve cells generate, receive, and impart?
10. How many varieties of nerve fibers?
11. Describe the function of the sensory fibers. The motory fibers.
12. What is a nerve?
13. How many membranes surround the brain.
14. Of what use are they?
15. Give the average weight of the brain of an adult.
16. Does a small brain necessarily indicate low intelligence?
17. Give a general rule about this. Any exceptions?
18. Into how many parts is the brain divided?
19. What is the name of the brain proper?
20. What is the function of the cerebrum?
21. How is the cerebrum divided? Is it completely divided?
22. Is the surface of the human brain smooth?
23. Do the convolutions vary in depth?
24. Do all animals have these convolutions? Name some that do not.
25. Is the gray or white matter on the surface of the brain?
26. What composes the bulk of the convolutions?
27. How do the convolutions affect the amount of gray matter?
28. Where is the cerebellum? Of what does it consist?
29. Where is the medulla? How large is it?
30. Give a general description of the spinal cord.
31. Where is its gray matter? Its arrangement resembles what?
32. How do the spinal nerves originate?
33. Which are the motory fibers? Which the sensory?

CHAPTER XXVI.

THE PHYSIOLOGY OF THE NERVOUS SYSTEM.

The Nerve Current. The peculiar force carried by the nerves from one part of the body to another is called the nerve current, or nerve force; it travels along the nerve at the rate of over one hundred feet a second. It is not known what this force is, although we do know what conditions favor it, and what check it. If the end of the finger be touched, almost instantly the sensation reaches the brain and is made known to the mind. But the nature of the force which conveys this sensation along the nerves from the finger to the spinal cord, and then up the cord to the brain, is, as we have said, entirely unknown.

The nerve current is set in operation by a stimulus. Suppose there is a desire to move the hand: in this case the stimulus is an act of the will, which excites the nerve current, so that it almost instantly passes down the nerve fibers to the muscles of the arm. When the nerve current reaches a muscle, the current itself acts as a stimulus, rousing the muscle to action. If we pinch the skin the stimulus is a mechanical one; the sensation is carried to the nerve centers, over the sensory fibers. The pupil of the eye becomes smaller when a bright light is brought near the face: in this case, light is the stimu-

lus which at last causes the muscle of the iris to contract.   A sharp scream will cause a person standing
near to jump; here fright is the stimulus.   Thus we
conclude that nerve currents, mechanical agents, light,
and the emotions, may each act as a stimulus.

Function of the Cerebrum.   The cerebrum is the seat
of the intelligence.   It is here that we think, feel, and
will.   It is not known how the mind is connected with
the brain.   It is impossible to explain how it is that our
intelligence depends upon the tissue of the brain.   We
simply know the fact that in some way the existence of
our mind depends upon a normal brain.   This has been
proved in many ways.   An injury to the head, resulting
in a portion of the skull being fractured and pressed
upon the brain, has been followed by loss of consciousness; and the person has remained in a deep sleep until
the surgeon has raised the depressed bone, when consciousness has returned.

Sickness has also proven it: an inflammation of the
membranes of the brain, affecting its surface, causes delirium and otherwise disturbs the mind.   Medicines
which affect the flow of blood to the brain also interfere with the reasoning faculties.   Persons born with
extremely small brains have no intelligence.   After removing the cerebrum all voluntary acts are abolished.
Thus injuries, disease, drugs, size, and experiments all
show that a healthy condition of the cerebrum is necessary for the existence of intelligence and the power
to will and to command.

Two Brains.   As stated in the preceding chapter, the
cerebrum is nearly divided into two complete parts.
This fact has led some physiologists to declare that

there are two brains, and that they act independently of each other. It is nearer the truth, probably, to say that so far as the mind is concerned, the sides of the cerebrum should be considered as one whole organ, but that each side controls the sensation and motion of the opposite side of the body.

Mind and Body. In some mysterious way, these two are so connected that what affects the one affects the other also. Therefore a healthy mind and a diseased brain cannot go together. As the same blood nourishes the brain that nourishes the other parts of the body, so the former cannot be in its best state unless the latter be in a healthy condition. From this it follows that if we would attain the fullest intellectual development, attention must be given to the laws of health, and their teachings strictly obeyed.

Function of the Cerebellum. Injuries to the cerebellum do not necessarily interfere with either the will or the consciousness, but with the movements of the body. From this it would appear that the cerebellum is especially concerned in maintaining a harmony of action of the voluntary muscles. Without the controlling power of the cerebellum, the muscles would act as irregularly as they do in one who is intoxicated. By its action we are able to keep an exact position ; and at all times can have the muscles act in harmony and with regularity.

Functions of the Medulla. This part of the nervous system is one of great interest. It is most essential to life, and it controls many of the most important functions. In the medulla are many " centers," or small collections of nerve cells, which control certain functions. When these centers are stimulated in any way, they put into action the functions they control; as for example, there is a " sneezing center." If some irritant be inhaled into the nose the ends of the nerve fibers are irritated, and an impression is conveyed to the sneezing center; from this center goes forth a nerve current to certain muscles, which contract and cause the expulsive act of sneezing. Besides the sneezing center there are many others, among which are: the coughing center; the center for the secretion of the saliva; the vomiting center; the swallowing center; and the center for the closure of the eyelids.

One of the most important centers is known as the respiratory center. It is a fact that a small collection of cells in the medulla controls absolutely all the movements of respiration. This small center has greater power than the will itself; for we may " will" not to breathe, and we can make the attempt to hold the breath, but soon we can do so no longer. Notwithstanding our greatest efforts, we again begin to breathe; for the center in the medulla is stronger than the will. We may be capable of increasing or diminishing the number of respirations per minute for a short time; even cease breathing for a brief period; but soon the respiratory center exerts its power, and respiration is continued with wonderful regularity.

Other important centers affect the movements of the heart. One center continuously holds the heart in check, causing it to beat with great regularity. Another center appears to have an opposite effect at times, being capable of accelerating the action of the heart.

No less important is the vaso-motor center. This controls the nerves which go to all the arterial system.

It is a small collection of cells, yet it is capable of causing the contraction, or relaxation, of the walls of any of
the arteries. It will be remembered that in the walls
of the arteries is a layer of involuntary muscle, arranged
in a circular manner around the vessels; if the muscle
contracts, the vessel will be narrowed; while if it relaxes, the vessel will be enlarged. The vaso-motor
center presides over the action of the muscular walls of
the arteries: the normal condition of this center is one
which keeps the arterial walls in a moderate state of
contraction at all times. The center is said to keep up
the "tone" of the arteries, whereby their walls are held
firmly in hand. When the function of this center is
checked, it releases its hold on the arterial walls and
they relax, thus enlarging the size of the vessel. If the
blood vessels are thus made larger, more blood will flow
through them, and the parts will be a deeper red in
color. This is usually temporary, but it may become
permanent.

Blushing is a temporary checking of the power of the
vaso-motor nerves of the face; they cease to act, thus
allowing the muscles in the smaller arteries of the skin
to relax; this results in a temporary enlargement of
the blood vessels, so that more blood flows to the part.
The vaso-motor nerves may become paralyzed, so that
the blood vessels are permanently enlarged; this is the
condition of the red nose of the confirmed inebriate.

The vaso-motor center may act in an opposite manner to that described, — it may act with more than normal power. The smaller arteries then have their muscular walls contracted with more vigor, so that the
amount of blood in a part may be greatly reduced.

Cold stimulates the vaso-motor center, so that if the hand be placed in cold water it soon becomes pale from loss of blood. Fright often causes the face to become pale, owing to the vigorous action of this center.

This power of the vaso-motor center is most essential to the preservation of health, and even of life itself. Let us illustrate its daily action: the cold weather of winter stimulates the center so that it acts with increased power; this contracts the arteries of the skin, so that the flow of blood through it is greatly diminished. Therefore the loss of animal heat is diminished, as we have already learned. But during the summer the vaso-motor nerves relax their hold on the smaller blood vessels of the skin; the blood flows more freely through it, and the loss of heat is thereby increased.

From all this we conclude that the vaso-motor center is capable of controlling the supply of blood to any part of the body. By increasing its normal function the arteries of any part are made smaller and the supply of blood correspondingly less; while by diminishing its normal work the arteries are made larger and the supply of blood increased. It has certainly been proven that the medulla is a most important, as well as most delicate part of the nervous system.

The Spinal Cord and Reflex Action. The spinal cord is the conducting medium between the nerve fibers of the lower part of the body, and the brain.

A second function of the cord is a reflex one. It is a great reflex center; its action in this respect is almost continuous. There are many familiar illustrations of this action in every-day life; tickling the foot of a person who is asleep causes the foot to be quickly withdrawn; this is purely a reflex act. The impression produced on the nerves of the foot is conveyed along the sensory fibers to the spinal cord, and from the cord it is "reflected" outward along a motory nerve to the muscles of the limb. The sensation produced by the tickling entered the cord through the posterior root of a spinal nerve, and immediately left it through the anterior root; this involved no interference of the brain.

To make a reflex act, three things are necessary : an unbroken sensory nerve, for connecting the point touched and the nerve center; a healthy nerve center; and an unbroken motory nerve, between the nerve center and the muscles to be stimulated. Reflex action is partly, but not altogether, under the control of the will. To illustrate : if we inhale an irritating powder, like pepper, through the nose, we may be able to postpone the sneezing for a short time, but finally we are obliged to sneeze, and no power of the will can prevent it.

Course of Nerve Current in Reflex Action. A glance at Fig. 98 will make clear the course of the nerve current in a reflex act. At the right, 3, is one of the spinal nerves. This large nerve consists of many fibers which proceed together until they reach some distant part, as the arm. Some of the fibers terminate in the skin, and others in the muscles. If the skin on the arm be touched the stimulus will be conveyed toward the spinal cord and will finally enter it through the sensory root, at 2. The nerve current then goes directly through the gray matter to the anterior or motory root, at 1 ; it then passes down the motory fibers, which are alone at 1, but which are soon side by side with the sensory fibers in the spinal nerve, at 3. After continuing the length of the

18

motory fibers the current finally stimulates some of the muscles of the arm and they respond by a vigorous contraction. Thus it is seen that the sensory and motory fibers are separated into distinct bundles at their beginning, and they are also separated at their termination, the former in the skin and the latter in the muscles; but they were together in one bundle through all the distance between. A reference to Fig. 99 may aid in making the subject more clear. The nerve current travels in the direction of the arrowheads.

Importance of Reflex Action. The daily work of the body is carried on largely as a result of reflex action.

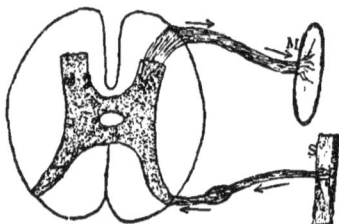

FIG. 99. A diagram illustrating reflex action : s, the skin, M, a muscle. If the skin, s, be touched the nerve current travels in the direction of the arrows until it stimulates the muscle, M, to contract.

The flow of saliva produced by mastication is a result of reflex action; and we could not check the flow if we desired. The flow of the gastric juice, from the stimulus of food in the stomach, is purely a result of reflex action. Respiration is a reflex act, due to certain stimuli applied to the respiratory center in the medulla. The nervous system is constantly performing a vast amount of labor of which we are unconscious, and which we are unable to alter, except possibly to a limited degree in a few instances.

Acquired Reflex Action. Many acts which are at first voluntary, and which are performed only by a strong effort of the will, finally become so natural and easy that they are performed unconsciously; these may be called acquired reflex acts. Walking was a very difficult process at first, and could only be accomplished with considerable effort. The number of muscles brought into play in running and jumping is very great; yet they all relax and contract at just the proper time, without the least aid from the will. When the beginner plays the piano, he not only looks at the music, but also at the keys, that only the right ones may be touched; but after a time the sight of a particular note calls forth such a movement of the hand that just the proper key is touched; the performer looks at the music, and the hands take care of themselves. The first trials at skating are not highly successful; all the power of the will has to be exercised to keep the balance and to move in the desired direction; but soon the movements become easier and the exercise becomes a pleasure and a rest.

Habit. A habit is an action acquired by frequent repetition. It is a law in physiology that each time a nerve cell acts in a particular way, it gains a power that makes the second act more easily performed. In this way a habit is formed: it may be the habit of walking, skating, or playing the piano; it matters not what the acts are, provided the nerve centers become accustomed to their repetition. At first, it was difficult to say the alphabet correctly, but now that we are in the habit of saying it, we can repeat it correctly more easily than incorrectly. It took an effort of the will at first to say it correctly, while now it requires much more will power to say it incorrectly; in other words, it takes a direct effort of the will to break the habit of saying it correctly.

This clearly proves two things: first, by the frequent repetition of an act we soon become quite unconscious of it; and second, it requires a direct effort of the will to break the habit. From this it follows that the nerve centers will have a strong tendency to do whatever they have been in the habit of doing. Thus there is sound reasoning in the statement that one kind act makes the way easier for another to follow. One duty performed makes another more easy; each pure thought invites another of its kind. Each temptation conquered makes us more sure of future victory. In youth, before the nerve centers have formed their life habits, it is comparatively easy for us to lay the foundation of future usefulness. Such, at least, is the teaching of physiology.

Life is a hard battle at its best, and if we would make it a grand success we need all the help we can get from every possible source. Physiology teaches us that by the repetition of what may seem trifling deeds, we can, while still young, form the habit of breaking the laws of health; it teaches us what the result of such actions will be; and it also teaches us that no person can avoid these results without avoiding their causes. On the other hand, we learn from this stern teacher that it is easy, while yet young, to form the habit of right living, so that we may build up both body and soul for future usefulness and happiness.

CHAPTER XXVII.

THE HYGIENE OF THE NERVOUS SYSTEM.

Necessity of a Healthy Nervous System. We have learned that all the functions of the body are dependent upon the activity of the nerve centers: these must be in the best condition, or the parts under their control will suffer. The nervous system is so closely related to the other parts of the body that an injury of any kind to the one is sure to affect the other.

Heredity. One of the things necessary for a vigorous nervous system is its inheritance. The old saying that "blood will tell" is as true to-day as when it was first uttered. We not only inherit the forms and the features of our parents, but also, to a certain degree, their dispositions, their powers of endurance, their likes and dislikes, and even their moral characters. The laws of heredity are very strong, and, in the long run, they are true. A child has a wonderful, almost inestimable gift, if endowed by birth with a sound and strong nervous system free from physical and moral ills. Such a child has a start in the world far in advance of others less highly favored, while his efforts to lead an upright life are much more easily crowned with success. The less favored child, who inherits a nervous system shattered and weakened by the vicious habits of his ancestors, enters life at a disadvantage, while he has a constant fight with himself to conquer a natural tendency to do evil. When at last his better nature rules, and he completely masters himself, he comes off a glorious conqueror.

A Healthy Body. Even with a vigorous nervous system and a strong moral nature transmitted to us, all our bright prospects may be easily blasted by neglecting to obey some of the well-known laws of health. To develop the nervous system requires proper food, pure air, and plenty of mental and physical exercise. Physical pain is not conducive to the best brain work; a tired body will not stimulate the brain to action; hence we conclude again that a healthy body is necessary for the highest and best development of the whole nervous system.

Exercise of the Mind. As muscular exercise is essential to the full development of the muscles, so mental exercise is necessary if the best efforts of the brain are desired. This exercise must be regular, persistent, and properly suited to the age and health of the person. It is impossible for the boy to jump so far at the first trial as he will after weeks of practice; neither can he perform so severe mental work at first as he will after months of constant study.

If the muscles have not been previously trained, one hard effort at lifting or running may cause severe pain and soreness in them. So an extra effort of the mind for one unaccustomed to study may cause headache and even severe mental disorders. The evils of over-study and of crowding too many studies into each school year are being more fully understood and corrected, while more attention is being given to a better understanding of the laws of health. The modern teacher knows that a pupil with a healthy body is much more likely to have a vigorous mind; and that the proper care of both should go hand in hand.

Mental labor ought not to cease with the school life; the school is to discipline and train the mind so that the powers of observation and reasoning may continue throughout life to the best possible advantage. " Work " is the word that contains the secret of a healthy body and a well-balanced mind, — regular, systematic, persistent work, both for the body and the mind. Pleasures are more enjoyable, and amusements are more profitable, if they be but the short vacations in our daily duties. A life of idleness and pleasure-seeking is not the normal condition of any human being.

Rest. The mind, as well as the body, would soon fail if it were obliged to work too hard or too long. Rest is absolutely necessary for all parts of the body. Many individuals seem to think that rest means to fold the hands and remain in perfect idleness; as a rule this is the poorest method of obtaining rest. If we have been exercising the mind until we are tired from study, nothing will restore the mental vigor better than some gentle out-door sport, as a brisk walk. The exercise brings a good supply of fresh blood to the brain, and thus aids in giving new life to its tissue.

The headaches of school children often cease before they have reached their homes at the close of day. Amusements, excursions, and a change in the character of the work, all tend to repair the waste of nervous energy, and are far better than idleness. Complete rest and quiet are not conducive to health unless for particular reasons they have been ordered by a physician.

Worry. Above all things do not worry. Study hard, play hard, enter with enthusiasm into all the duties and pleasures of school life, but do not worry. Worry means waste, — waste of nervous force, of thought, of memory ; and it is a sure road to the impairment of the highest functions of the brain. While it is true that some students do not have anxiety enough for their work, yet it is equally true, especially in the higher grades, that many attempt to accomplish altogether too much.

Sleep. One of the great restorers of both mind and body is sleep. All animals having a well-developed nervous system take rest in sleep. Drowsiness and weariness warn us that sleep is necessary. These warnings may be unheeded for a time, but sooner or later we have to yield to the imperative demand. Some persons require more sleep than others, but the adult needs, on the average, from seven to nine hours. It is said that Napoleon required but three or four hours sleep each day, and that he would pass days with very little rest of any kind. Frederick the Great required but little sleep, not over five hours a day. These are marked exceptions, though it is a fact that nearly all our great men who are obliged to do an immense amount of brain work sleep well and long; they know the value of a good night's rest, and are alarmed when they are unable to procure sleep. They know that during their busy days the waste is greater than the repair, and that during the quiet rest of the night the cells are busy repairing the waste, and appropriating new material for the labor of another day.

Insomnia. Continued wakefulness often becomes a very serious affection, and many men have lost their health by the inability to sleep. If persistent, it is a dangerous trouble and should be remedied at once, if possible.

To promote Sleep. There are many popular ways of promoting sleep, nearly all of which make a bad matter worse. Out-door exercise during the day, light suppers, quiet evenings, and warm feet will greatly promote sleep. " Keep the head cool and the feet warm." But sleep will often refuse to come if the tired brain is filled with cares, griefs, and anxieties.

Students need Much Sleep. Students often make a great mistake in trying to change the laws of nature, and in studying until late into the night, and then in sleeping away the morning hours. It is only necessary to remember that at night the brain is tired, and an extra effort is necessary to make an impression upon it ; while in the morning it is fresh and sensitive and easily impressed ; the morning is the time for study. Earnest application in the early part of the day, concentrating the mind with all the power of the will, and laying aside everything else but the work in hand, — these will soon prove that the mornings were made for study, the evenings for relaxation, and the nights for sleep.

QUESTIONS.

1. What do you understand by the "nerve current?"
2. How fast does it travel?
3. What sets the nerve current in operation? Illustrate this.
4. How has it been proved that the cerebrum is the seat of the intelligence?
5. Do injuries to the cerebellum necessarily destroy consciousness?

6. What is the especial function of the cerebellum?

7. What is said of the medulla in its relation to life?

8. Name some of the centers in the medulla.

9. What controls the movements of respiration? Illustrate this.

10. What centers affect the movements of the heart?

11. What center controls the nerves which are in the walls of the arteries?

12. What is the normal condition of this vaso-motor center?

13. When the function of this center is checked, what occurs?

14. Give the physiology of blushing.

15. When the vaso-motor center acts with more power than the normal, what occurs?

16. What is this center capable of doing?

17. The spinal cord is a connecting medium between what?

18. Give another function of the cord. How illustrated?

19. What three things are necessary for a reflex act?

20. Is reflex action ever under the control of the will? How illustrated?

21. Describe the course of the nerve current in a reflex act.

22. Give some illustrations showing the importance of reflex action.

23. Are reflex acts ever acquired? Give illustrations.

24. Give a definition of a habit.

25. Give some illustrations showing that by frequent repetition of an act we may become unconscious of it.

CHAPTER XXVIII.

ALCOHOL AND THE NERVOUS SYSTEM.

AFTER the habit of indulgence in alcoholic drinks has been fully established it has such a mastery over the whole system that statements or warnings of any kind have little effect. But it is doubtless true that if all young men knew the effect of alcohol on the nervous system, before they began to use it, few, if any, would ever try the experiment of taking the first glass ; while it is certainly as true that no right-minded young woman, possessed of a like knowledge, would ever think of offering a glass to a friend.

It is because of the ignorance of the terrible power hidden in strong drink that young men ever begin the use of alcohol as a beverage. " One glass certainly cannot do me any harm," is the oft repeated statement. The student of physiology knows full well, though, that even the one glass, repeated now and then, has the power to make such changes in the tissues and organs of the body, and to produce such a marked effect on the nerve centers that breaking away from it will be very difficult. And some day it will be discovered, when too late to retreat, that a habit has become fastened with wonderful tenacity.

There are two classes of persons who begin the use of

alcoholic beverages : those who honestly think there may
be some healing virtue in their power, and therefore take
some form of alcohol as a tonic ; and those who do so
for no particular reason, only because they are asked.
The persons of the first class are deceived by the primary narcotic effects of the alcohol, which by deadening
the nerves makes them feel better for a time, and so the
doses are repeated until the habit is formed.

Many physicians have erred in prescribing some form
of alcohol as a regular medicine, under the mistaken
impression that it is a food and tonic. All at once
the patient discovers he has a most unnatural desire
for his medicine, and is powerless to give it up. To
such as believe that there is virtue in alcohol as a tonic,
and that its use will be of benefit to them in restoring
health, we can only say : Never prescribe it for yourself
any more than you would prescribe opium or strychnine,
or any other poison.

The wise physician of to-day, who is abreast with the
modern investigations concerning this drug, knows how
little virtue there is in the various forms of alcoholic
beverages as a real aid in combating disease. Doubtless a great deal of alcohol used as medicine comes from
the self-prescribing of those who take it because they like
it. A better understanding of its physiological action
has made it necessary to regard alcohol as a dangerous
poison. Some of the most competent medical men in
this country have openly declared that for many years
they have never ordered a single dose of alcohol in any
form, and that there are better and safer remedies which
can be given in its place.

To the second class belong those who are beginning

the use of alcohol because some one offers to treat them
and they do not wish to refuse. To these, and also to
all who have not ventured thus far, we would say :
Study carefully the following effects of alcohol on the
nervous system, and then calmly decide whether the
path is not altogether too treacherous and dangerous
to enter upon.

The most striking thing about alcohol is that *it has a
special affinity for nervous tissue.* By " affinity " is
understood the attraction which takes place between the
alcohol and the tissues of the body. When alcohol is
taken into the body it affects all the living tissues, but
it affects them in different degrees ; it has some affinity,
or attraction, for muscles, some for the blood, some for
the liver, and so on ; but of all the tissues in the body
it has the strongest affinity for those of the nervous system. Thus, if after death from the effects of alcohol,
the tissues are separately analyzed by the chemist, more
alcohol will be found in the brain and spinal cord than
in any other tissue.

The brain is the throne of the intellect, and yet it is
attacked the most severely of any of the tissues. Where
man is strongest, there does alcohol make its most
powerful attempt to overthrow him. Many a simple
drug can do great damage in time by persistently at-
tacking one of the less important tissues. Many diseases
affect the less important organs and tissues at first, and
it is only after years of annoyance that the general
health becomes undermined. But alcohol strikes at
once with its full power at that which is highest and
most important to man.

In the preceding chapter it is taught that the nervous

system controls all the functions of the body ; anything which affects the governing body must affect the parts governed. So we find alcohol affecting with the greatest severity that which is the most central and vital. Its affinity for nervous tissue is so great that it has been found in the fluid normally present in the brain, and it has even been distilled from the brains of those who have died after using large quantities of it.

When alcohol is taken into the body in the form of some kind of drink, it is absorbed into the circulation and is carried to all the tissues of the body, more going to the nervous tissue than to any other. Its effect on this tissue is that which must follow from the administration of any powerful narcotic. Alcohol paralyzes nerve tissue. A standard medical work on the action of the various drugs on the human system, classifies alcohol with " those remedies which diminish or suspend the functions of the cerebrum after a preliminary stage of excitement." Webster says that paralysis is the complete or partial abolition of function. Therefore we are correct in stating that alcohol paralyzes nerve tissue; for it completely or partially abolishes the functions of the nervous system.

A standard writer on materia medica says that the effects of alcohol are expended chiefly on the nervous system ; this fact alone is enough to condemn its use in any form as a beverage. But it makes itself appear to good advantage before those who do not know its full effects, because of its power to first excite the brain. We understand the cause of this excitement when we know that alcohol paralyzes, and so checks the action of the vaso-motor center, causing the latter to release its hold over the muscular walls of the small arteries. As a result the vessels expand and more blood flows through them, and following this enlargement of these smaller blood vessels the distant pressure is removed and the heart beats faster. The more rapid flow of blood and the enlarged blood vessels bring an extra supply of blood to the brain and thereby it becomes greatly excited. The flushed face is the result of the same changes; the walls of the smaller arteries are relaxed and the vessels become filled with blood, thus causing the skin to appear red. In the light of these facts we must regard the flushed face and the excitement of the brain as symptoms of a temporary paralysis of the nerves which control the circulation of the blood in the smaller arteries.

Following the brief period of excitement produced by alcohol, other effects appear. The spinal cord becomes involved, as shown by the irregular action of the parts under its more immediate control. It is unable to fully perform all reflex acts, and the will has to come to its aid. Thus the act of walking becomes uncertain, and it requires the aid of the will to order the movements of the proper muscles. The loss of some of the power of the spinal cord is also shown in the trembling of the lips and the uncertain and indistinct speech. A little later the cord loses much of its force as a nerve center, and its control over the muscles is slight; then the muscles of the lower limbs become weak and the gait is staggering. All this means a temporary paralysis of the nerve centers. The centers in the medulla soon become affected, and vomiting is likely to occur; this may prove beneficial by throwing off the poison, but usually the alcohol has been all absorbed from the stomach before vomiting takes place.

When the stage of excitement, or increased action of the brain cells, is at its height, the cerebrum acts in a still more excited manner; the nerve cells become over-excited, and the animal nature of man assumes control; the reason is overpowered and man is brought to a level with the lower animals. The muscles are no·longer under control, and the whole system seems to be ruled by some strange outside power. The stage of excitement is soon over, and gradually the entire brain loses its power; the voluntary muscles no longer act; all sensation is lost; and the body becomes nothing more than a wonderful piece of disordered mechanism. "The functions of the cerebrum are suspended."

Yet the centers in the medulla are still active, for the heart continues to beat and the respiratory muscles continue to act. These two centers would soon become paralyzed also, and the heart and lungs would cease to do their work, were it not that by reason of unconsciousness the person takes no more alcohol. Herein lies the explanation why alcoholic drinks do not cause more sudden deaths; the brain loses its power before the heart ceases to beat. In other words, the poison affects the cerebrum before it does the medulla. The man becomes unconscious before he has taken enough to check the action of all the centers in the medulla.

If alcohol affected the nerves which keep the heart in motion before it did the consciousness, it would cause a vast number of sudden deaths. But it may be urged that this is an over-drawn picture; that it is an illustration of drinking to complete intoxication, and that even those addicted to the use of alcohol do not favor such excesses. We must, therefore, study the effects of moderate and frequently repeated doses of alcohol on the nervous system.

We must fully understand that alcohol has an immediate and direct effect on nerve tissue. By direct effect is understood the effect produced by actual contact of the alcohol with the nerve tissues; for some of the alcohol taken into the system comes in direct contact with the brain, producing most marked changes in the structure of that organ. This statement has been proven by experiments, by chemical analysis, and by microscopical examinations, — sustained by the reports of eminent men. The structure of the brain is also changed, indirectly, by the poorer quality of the blood, by the impaired circulation, and by the diminished activity of the organs of excretion; thus the supply of blood is less, the blood itself is poorer in quality, and the channels for the removal of the waste and worn-out materials are not sufficiently active.

As a result of both the direct and indirect effects of alcohol the following marked changes are produced. An eminent medical authority says that after the continued use of alcohol " the nerve cells of the gray matter are more or less fatty and shrunken." As a result of this shrinkage he says, " The whole cerebrum becomes smaller, and the space thus made becomes filled with a watery fluid." It is impossible to place too much emphasis on this statement. This was written to physicians and medical students as a statement of the facts concerning the action of one of the many drugs used in the practice of medicine.

19

What is the teaching of this high authority? That the continued use of alcohol actually lessens the size of the brain; reduces the capacity of man to think; changes the mysterious and wonderful nerve cells so that they are smaller and partly changed to fat. Space once occupied by brain tissue becomes filled with water. Thus, in this cool and logical way, are we brought face to face with the astounding anatomical fact, that alcohol has the power to "steal away the brains." "The evidences of these changes in the brain and cord," the same authority says, "are seen in the impaired mental power, the muscular trembling, and the shambling gait of the drunkard."

It would seem to any logical mind that nothing more need be said to prove the baneful effects of strong drink. The knowledge of the system already acquired from a study of the preceding chapters, must be sufficient to show how surely alcohol will do its work, and also how severe that work will be.

But testimony accumulates on every hand. Dr. William A. Hammond, the eminent specialist on nervous diseases, made a careful study of the effects of alcohol on the nervous system. If it had any special virtue he was desirous of employing it in his practice. By numerous experiments he proved its powerful poisoning effects. He says: "Mankind would be better mentally, morally, and physically, if the use of alcohol were altogether abolished."

Delirium Tremens. This is one of the horrible effects of alcoholic poisoning. It may follow the sudden use of a large quantity of alcohol, even from a single intoxication, although it is usually the result of the free and long continued use of alcoholic drinks. There is no telling at what time it may attack its victim; it may be when he first begins his career, or after years of continued drinking; it may come to those physically strong, or to the very weakest. It is one of the most horrible effects of alcohol, transforming man into the wildest and fiercest of animals. There is inability to retain food and all the voluntary muscles are in a tremor; the patient is either unable to procure sleep at all, or there are brief periods of sleep, interrupted by the most terrifying dreams; he is being constantly pursued by horrible insects and reptiles, while he is unable to escape from them, or to destroy them. Recovery may take place after days of intense agony, or death may bring relief at any moment.

THE EFFECT ON THE MIND.

From the preceding statements it logically follows that the use of alcohol diminishes the will power. Men who are proud of their dress and speech, and who pride themselves on their proper decorum, become foolish and silly when slightly intoxicated, and are easily provoked to do rash things. A small boy can so provoke a slightly intoxicated man by imitating him that the man appears to lose all control of himself. Even a small amount of strong drink is sufficient to make some people more easily annoyed and disturbed than when perfectly free from the effects of alcohol. Thus we reason that the man who indulges in strong drink, but not to the extent of intoxication, does not have full control of his mental and moral powers. He is more easily provoked and

more readily yields to temptation. Man needs all the
will power he can command, to successfully meet the
many trials and battles of life. To take strong drink
is to weaken his mental power, and by so much to diminish his chances of success in any of the useful activi-
ties of life.

To the conclusions of logic and the positive statements
of medical authority as to the undermining and weakening of the mind by alcohol, there may be added the testimony gathered from the records of courts and prisons.
Statistics show that a large percentage of the various
crimes are committed while the persons are under the
influence of liquor. Alcohol blunts all the finer sensibilities, and destroys the love for justice and fair play.
The mind is trampled upon and the lower animal nature
rules the body. As a result, deeds are committed which
would never be considered for a moment if the higher
nature were in command ; or, in other words, if the king
were on his throne.

Other diseases of the nervous system result from the
use of alcoholic beverages. The statistics of the insane
asylums show that insanity is one of these results.
What can be more terrible than to see a promising
career end in the loss of reason, and to find that strange
and wild fancies occupy its place ? In France, during
the last war with Prussia, it was found that over one-
half of all the cases of insanity were occasioned by the
use of alcohol. In the lunatic asylum at Dublin, " nearly
one half of the cases were known to be caused by the use
of alcohol alone." In America the proportion is not so
great, yet it is surprisingly large. It is safe to assert
that of one hundred insane persons, twenty are so afflicted

as a direct result of the use of alcohol, and thirty-five more as an indirect result, making in all fifty-five of each one hundred insane persons so afflicted on account of the use of alcohol in some form.

Still another effect is seen in the weak nervous systems transmitted by drinking parents to their children. Some form of nervous disease is very generally inherited from parents addicted to strong drink; this transmission makes their children less able to withstand the various diseases of childhood. The number of children who die from weak constitutions, inherited from intemperate parents, is very large indeed. With stronger constitutions they would be able to survive many of the ailments incident to childhood, but they appear to have no power to resist disease. One of the prominent causes of infant mortality in our large cities, if the truth were only told, would be inherited alcoholic poisoning.

The children of intemperate parents are likely to grow up weak in body and mind. Some go to the idiot asylums, some to the jails and prisons, while some go to the retreats for the insane. Others, however, manfully fight the battle; though tempted as none others are tempted, they conquer; though naturally weak, they develop strength; and though they inherit a love for strong drink, they persistently fight for its overthrow.

No battlefield has records of victories more noble. When these triumphs are gained, the tide of evil is turned back forever; a fine mental organism and great will power, the constituents of greatness, are gained; and the person who inherited weakness is now a source of strength to all who are associated with him.

OPIUM.

Opium is a most powerful drug, and is classed with alcohol in its power, first, to excite, and then suspend the functions of the cerebrum. Its effects are more pronounced than those of tobacco, and are fully as severe as those of alcohol. In this country the opium habit is generally formed as a result of the use of opium to relieve pain. The individual has usually been a great sufferer and the physicians have been obliged to prescribe some form of opium to give relief. From these prescriptions for relief from pain the patient has sometimes been left with a strong desire for more opium, and by repeatedly yielding to the desire the habit has been formed.

The effects of opium on the nervous system are most pronounced; it causes a partial paralysis of the lower limbs, giving a stooping or creeping appearance; it so interferes with all the functions of the body that the nerve centers suffer greatly from the want of nourishment. Opium has a wonderful power to blunt the moral sensibility; the opium-eater will do anything to obtain the drug; the mind soon becomes enveloped in a cloud, and he goes about in a dazed stupor. All interest in business and friends is lost, and the one dominant idea is how to procure more of the drug; to do this he will almost invariably deceive in every possible way. If he is unable to get his accustomed dose he suffers the most terrible pains both in mind and body.

QUESTIONS.

1. What knowledge would probably keep young men from beginning the use of alcohol ?

2. What two classes of persons begin the use of alcoholic beverages ?

3. How are those of the first class deceived ?

4. How must we regard alcohol ?

5. What do some competent medical men declare ?

6. What advice is given to the second class of young men ?

7. What is the most striking thing about alcohol ?

8. What do you understand by affinity ?

9. What tissue is attacked most severely by alcohol ?

10. How does alcohol affect nervous tissue ?

11. With what remedies is alcohol classed ?

12. Give a definition of paralysis.

13. Where are the effects of alcohol chiefly expended ?

14. How does it make itself appear to good advantage ?

15. What is the result of checking the action of the vaso-motor center ?

16. What effect does the extra supply of blood have on the brain ?

17. What follows the brief period of excitement produced by alcohol ?

18. What do the trembling lips and the uncertain speech show ?

19. When the stage of excitement is at its height, how does the cerebrum act ?

20. How is the reason affected ?

21. Describe the condition after the stage of excitement is over.

22. Why do not alcoholic drinks cause a greater number of sudden deaths ?

23. When taken into the system, does alcohol ever come in direct contact with the brain ?

24. How does alcohol affect the brain ?

25. How does it affect the will power ?

26. What do the statistics of insane asylums show ?

CHAPTER XXIX.

THE SENSE OF SIGHT.

Protection for the Eyes. The eyes are well protected in deep sockets of bone, called the orbits. Externally they are protected by the eyebrows, the eyelids with their glands, and the lachrymal glands. The nose is also a valuable protection to the eyes.

The eyebrows project over the eyes and are covered with a thick growth of hair. The hair is directed obliquely outwards, so that the perspiration from the forehead is carried to the side of the face, instead of running directly down into the eyes.

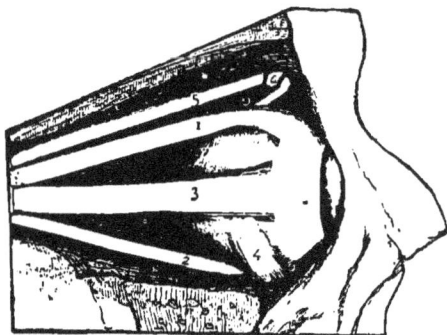

FIG. 100. The muscles of the right eyeball. The outer bony walls of the orbit have been removed. (1) the muscle which turns the eyeball upward; (2) downward; (3) outward; a corresponding muscle on the inner side moves the eyeball inward; (4, 5,) muscle which rotate the eyeball; (6) a pulley, through which the tendon of the muscle (5) moves.

The eyelids are thin curtains placed directly in front of the eyeballs. In the center of each eyelid is a thin plate of cartilage, on the outside of which is a thin muscle covered with skin. The inside of the lids is lined with a delicate membrane, called the conjunctiva.

On the edges of the lids is a row of delicate hairs, called eyelashes. They protect the eyes from insects and particles of foreign matter; for the moment any foreign bodies come in contact with the lashes the lids close, thus preventing the objects from touching the eyeball. The eyelids keep the heat and cold from the more delicate parts of the eye, and they also keep out an excess of light. Their most important function is to cleanse the eyes and to keep them moist. Their rapid and frequent movements thoroughly remove any particles of dust from the front of the eyeball, while at the same time they moisten the surface. This is the object of winking, which is usually a reflex act, although it may be made voluntary.

The Oil Glands. Oil glands are situated at right angles to the free edges of the lids, and on their inner side. They can be seen on the inner surface of the eyelids through the mucous membrane, looking like strings of minute pearls. They extend nearly across the entire width of the lids, on the edges of which they open and pour out their oily secretion.

FIG. 101. The eyelids of the right eye, viewed from the inside: (1) the lachrymal gland; (2) the oil glands in the eyelids.

This keeps the lids from adhering to each other, and holds back the tears so that they do not run over the edges and down upon the face.

The Lachrymal Gland. The lachrymal gland is an almond-shaped body situated in the outer and upper part of each orbit, between the eyeball and the bone. From this gland there extend about seven ducts or canals, which open on the inner surface of the upper eyelid near its outer part. The openings are arranged in a row, as represented at 2, Fig. 102, thus distributing the secretion over the surface of the conjunctiva.

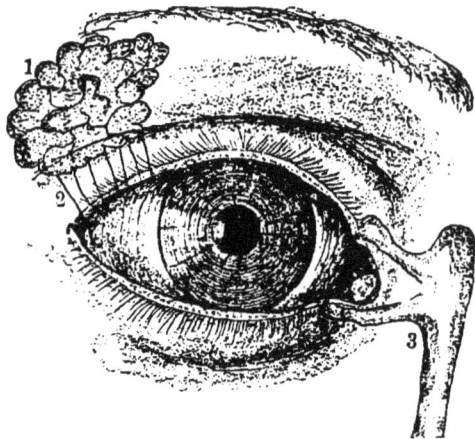

The Tears. The watery secretion from the lachrymal glands is known as The lachrymal fluid. The secretion is a con-

FIG. 102. Front view of the right eye, showing the location of the lachrymal gland and the nasal duct: (1) the lachrymal gland ; (2) the ducts that carry the secretion from the gland to the free surface of the eyeball ; (3) the duct for the passage of the secretion to the nose.

stant one, although we are unconscious of its presence until there is an excessive flow ; the fluid is evenly distributed by the movements of the eyelids. An excessive amount of this secretion is called the tears. They are easily excited by irritants placed to the eye or nose ; by laughing and crying ; and by various mental emotions.

Some of the secretion is evaporated from the eyeball, but the greater part of it escapes from the eye through regular channels provided for it. The secretion flows toward the inner angle of the eye where it enters two openings, one in either lid. This opening is easily seen on the lower lid as a black point in the center of a little eminence near the inner corner of the eye. The black points are the beginnings of two ducts which pass inward toward the nose, as seen at 3, Fig. 102.

The Eyeball. The eyeball is securely protected from injury and yet it has a most extensive range of sight. It is a round body with the exception that its front part protrudes more than the other parts. From the front to the back it is about an inch in length.

The eyeball has three membranes or coats surrounding it. The outer coat consists of two parts; the posterior part, which is very thick and hard, is seen coming toward the front of the eye where it is called " the white of the eye." Because it is so very tough and hard, it is called the sclerotic; the anterior part of this outer coat is transparent and is called the cornea. Through this transparent cornea the light passes into the interior of the eye.

The middle coat of the eye is called the choroid; it is filled with dark pigment which makes it look quite black. The choroid joins a membrane in front, called the iris, in the center of which is an opening, called the pupil. In the iris are the cells, containing a coloring matter, which give the characteristic color to the eye. The size of the pupil can be changed by the action of certain muscles: the muscles are involuntary so that they only contract and relax as a result of some influence outside of the will. A bright light will make the muscles of the iris contract so that the pupil becomes much smaller, and thus but little light reaches the interior of the eye. If the light be very faint, other muscles of the iris contract, enlarging the pupil, and thus allowing more light to enter.

The third, or inner coat of the eye is called the **retina.**

The miscroscope shows that it has a most complicated structure, and that it is directly connected with the fibers of the optic nerve, which go directly to the brain. When light reaches the interior of the eye, it

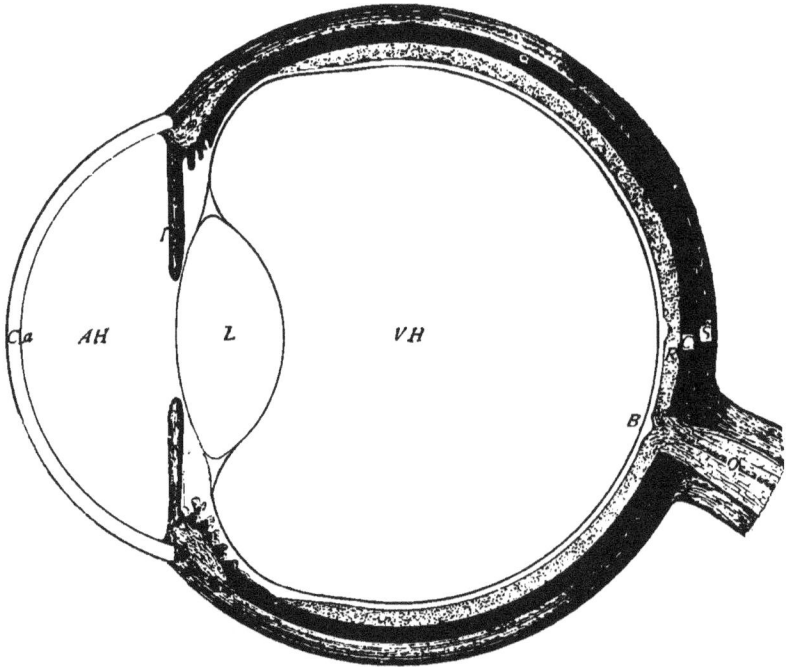

FIG. 103. A section through the eyeball : Ca, the cornea, or the transparent membrane which forms the front of the eye ; s, the sclerotic ; c, the choroid ; R, the retina ; o, the optic nerve ; B, the blind spot ; I, the iris ; L, the lens ; AH, the aqueous humor ; VH, the vitreous humor.

produces a peculiar impression on the retina ; which impression is then conveyed along the optic nerve to the brain ; the brain transforms the impression made upon it into the sensation of vision, and we declare we are able to see. When any object is viewed the exact image of it is produced on the retina. If the optic nerve be cut the image will still be formed, but no sensation of light will reach the brain. Therefore, as it is the photographer who sees and not his camera, so it is the brain which sees and not the eye, for the eye is simply the camera.

Functions of Parts. The eyeballs may be likened to a room, with a single window in front. Just back of the window hangs a dark curtain with a round opening in it. All light entering the room must come through the window, pass through the opening in the curtain, and strike the opposite wall. The transparent cornea is the window; the iris is the dark curtain for regulating the amount of light; and the pupil is the central opening. The choroid is black to prevent the reflection of light within the eye, and to absorb any light which may pass through the retina. The sclerotic is hard and firm, for a protection to the eye and for the attachment of muscles.

The Blind Spot. All parts of the retina are not equally sensitive to light. One spot on it, where the optic nerve

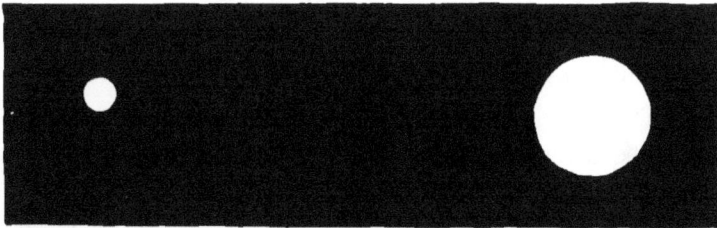

FIG. 104. A diagram for illustrating the existence of the blind spot.

enters (see Fig. 103, B), is entirely insensible to light. This is called the blind spot: it does not interfere with vision because it is impossible for the light from an object to fall on the blind spot of both eyes at once. If certain rays fall on the blind spot of one eye, they will fall on a different part of the retina of the other eye. But if one eye be closed there is always some portion of the object before us which is invisible. This is easily proved by looking at a sharply defined object after the following manner: Close the left eye and look steadily at the small white circle to the left of Fig. 104: it is possible now to see the large white circle even when the eye is fixed on the smaller one. Hold the book vertically on a level with the eyes at a convenient distance. Now move the book slowly backward and forward. Soon a distance will be found where the large circle entirely disappears, only to reappear again as the book is moved nearer or farther from the eye. This is because the light from the large circle, when it entered the eye at a certain angle, fell on the retina just where the optic nerve enters.

Care of the Eyes. The sense of sight is one of the greatest of gifts. If a man has a sound mind so that he is able to understand and appreciate what is brought to his notice, great pleasure and profit can come to him through the eyes alone. But great care is necessary if we intend to keep the eyes in their best condition. Many eyes are harmed, and some ruined, by their improper use during school life. The conjunctiva is easily inflamed from a cold or some other cause. In such cases a physician should always be consulted, that the trouble may be relieved at once before it becomes chronic.

The eyes may look bright and clear, yet if their use in reading is followed by pain in the head, it is probable that there is some defect in vision. This condition should not be neglected, but the exact cause of the trouble should be discovered and the proper remedy applied. A clear and steady light is most desirable; a dim light makes it necessary to put forth an effort to see, while a strong light is equally as injurious. Looking at a brilliant light, as the sun, is positively injurious. The light should come from over the shoulder, so that it may fall on the page of the book without coming directly into the face. Squinting, or looking cross-eyed, or rolling the eyes about, as is often done by children in sport, is a dangerous thing to do, as some of the muscles of the eyeball may be severely strained by so doing.

The upright position is the natural and proper one for reading. Reading while lying down, especially in bed, is a very unwise practice. The book should be held from twelve to sixteen inches from the face when reading. Those who cannot see ordinary type, unless the book is nearer than ten inches, should consult an oculist and ascertain if it would not be best to wear glasses.

The eyes should have complete rest if they are inflamed, and eye-washes and ointments should not be used without the advice of a physician. Never rub the eyes, for any reason, especially if some particle of dirt has fallen into them; have all such objects carefully removed at once by some competent person.

The best general advice that can be given is this: As soon as the slightest failure of vision occurs, or when anything unusual is noticed about the eyes, do not wait for them to cure themselves, but seek the best medical aid without delay.

ALCOHOL AND THE EYES.

The bleared eyes of the hard drinker only too clearly show the effects of this poison on the delicate tissues of the eye. The blood vessels of the conjunctiva, entirely invisible in health, become distended to their utmost with blood, giving the eyes a most undesirable appearance. Here the cause is the same as has been previously described. The vaso-motor center, in the medulla, is so affected that the vaso-motor nerves release their control over the smaller arteries; the muscular walls relax; and the vessels become engorged with blood. If they remain in this condition too long they will not return to their former size, even if no more alcohol be taken. But if the habit be discontinued early, it is possible under proper medical treatment to restore the parts to health. Inflammation of the eyes is only one of the many serious effects of alcohol on the various organs and tissues of the body. In a recent treatise on diseases of the eye this statement occurs: " It is a well-established fact that long continued and frequent indulgence in small quantities of spirits is very deleterious to the eyesight. The persistent morning nausea, muscular tremors, sleeplessness, and dull headaches that plague the chronic drinker, are likely to be associated with degeneration of the optic nerve tissues."

A very noted oculist examined the eyes of one thousand persons who were known to be addicted to the use of strong drink. He reported that of these one thousand cases, three hundred had eye affections of some kind. In one hundred and eighty cases the affections were of a severe nature. He gave it as his opinion that alcohol was the cause of the disease in each of these cases.

TOBACCO AND THE EYES.

The edges of a smoker's eyelids are often inflamed as a result of contact with the irritating smoke. Often the smoker experiences sharp pains in the eyeballs, with slight failure of vision.

The most serious results to the eyes sometimes follow the use of tobacco; one of these is known as "smokers' blindness." The most careful examinations of the eye fail to discover any change in its structure, yet gradually the vision becomes less and less until complete blindness results. We have the testimony of a number of eminent oculists to the effect that persistent smoking by the young causes frequent and severe affections of the eyes.

QUESTIONS.

1. How are the eyes protected?
2. Describe the eyelids.
3. What membrane lines them?
4. Of what use are the eyelids?   .
5. What is the object of the oil glands?
6. Where are the lachrymal glands?
7. What outlet is provided for the lachrymal fluid?
8. How many membranes surround the eyeball?
9. Name the posterior part of the outer membrane. The anterior part.
10. Where is the choroid? The iris? The pupil?
11. What is the inner coat of the eye called?
12. With what is the retina connected?
13. Why is the choroid black? Why is the sclerotic hard?
14. Where is the blind spot of the eye?
15. Look at Fig. 104, as suggested on page 302.

20

CHAPTER XXX.

THE SENSES OF TASTE AND SMELL.

The Tongue. The sense of taste is located in the tongue, the back part of the roof of the mouth, and, to a slight extent, in the sides of the throat. The mucous membrane of the tongue, however, is more especially the seat of this sense. The tongue is composed of voluntary muscle, covered with mucous membrane. In health, it is moist and of a light-red color. Any marked change from this condition is an indication of some departure from health. Thus the appearance of the tongue often gives aid to the physician in ascertaining the source and character of the disease.

Papillæ of the Tongue. The mucous membrane of the tongue is covered with a great number of papillæ. In these papillæ are found blood vessels, nerves, glands, and " taste buds." There are three varieties of papillæ on the surface of the tongue. The largest papillæ can be seen by the unaided eye ; they are far back at the base of the tongue, arranged in the form of the letter V, with the point of the V toward the back. There are eight or ten of these, each consisting of a central papilla surrounded by a wall. The second variety can also be seen with the unaided eye, scattered freely over the surface of the tongue. These papillæ are easily

recognized by their deep-red color, and because they are
larger than the third variety.   They are most abundant
at the tip of the tongue, where they present a club-shaped

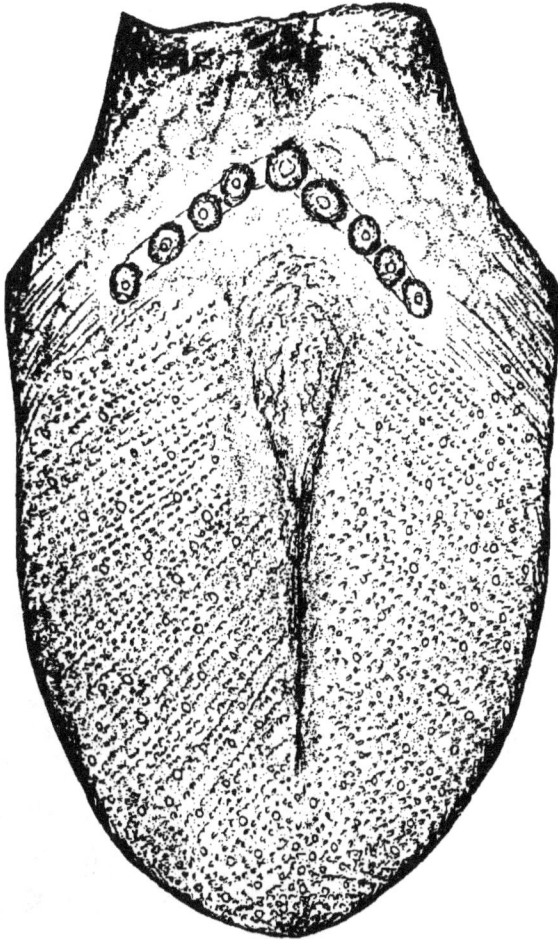

FIG. 105.   The tongue, showing the varieties of papillæ.

appearance.   The third variety is the most numerous
of all.   The papillæ are minute in size and are evenly distributed over the surface of the tongue.   They are of a
whitish color, owing to their thick epithelial covering.

In some of the papillæ are found loops, or coils, of minute blood vessels; in others there is the ending of a nerve fiber, giving the sense of touch to the tongue; while in the other papillæ are minute bodies, especially for the sense of taste.

The Taste Buds. The taste buds are found in the large papillæ at the base of the tongue; and a few are also distributed to other papillæ. They are collections of cells arranged in the form of buds, hence called taste buds. Each bud is not over $\frac{1}{300}$ of an inch in length. The location of these bodies,

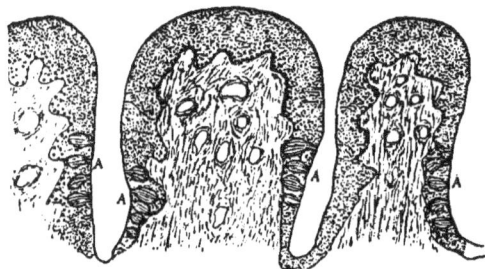

FIG. 106. A section through two papillæ of the tongue, showing the taste buds, at A, magnified.

in the edges of the papillæ, is illustrated in Fig. 106. It is seen that they are situated in the folds between the papillæ, rather than on the upper free surface. Fig. 107 shows some of these bodies more highly magnified. Some of the cells composing each bud are directly connected

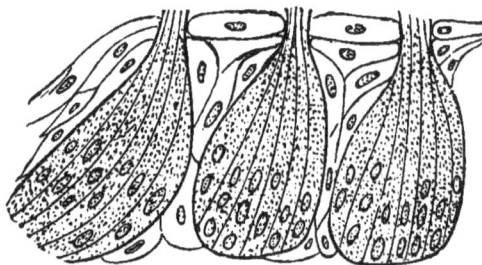

FIG. 107. Three of the taste buds of Fig. 106, highly magnified.

with a nerve fiber so that whenever anything comes in contact with these cells an impression of its "taste" is conveyed down the cells and along the nerve fibers to the brain.

The Sense of Taste. There are four different qualities of taste. We have the sensations of sweet, bitter, acid,. and saline. In order that any of these may be appreciated, the substance must be dissolved. Dry sugar placed on a perfectly dry tongue produces no sensation of sweetness. Some of it must be dissolved before any effect is produced on the cells of the taste buds. The saliva aids in this, although there are mucous and serous glands in the tongue which secrete a watery fluid. The movements of the tongue promote the flow of these secretions, and thus aid in dissolving the substances and in distributing them over a greater surface. The sense of taste can be greatly improved by practice, while it is materially aided by the sense of smell.

Confusion of Taste and Smell. The senses of smell and taste are often confused. Many times we believe we taste a substance when it is only the odor which is perceived. It is stated that neither vanilla nor garlic has any taste; it is their odor alone which is noticed. The odor of a drug is often more disagreeable than its taste; for this reason many medicines are best taken after first closing the nose and thus shutting off the odor. A severe cold is said to affect the sense of taste; this is because the lining membrane of the nose is inflamed, and we are unable to distinguish odors. In man the sense of taste is more highly developed than that of smell, while in some of the lower animals, as the dog, the sense of smell is the more acute.

Taste Easily Changed. The taste of many substances which were at first very pleasant may become disagreeable because of too frequent use, or of unpleasant associations; many articles of food are distasteful when first used, but after a time they are greatly desired. Many persons have had to make repeated trials before becoming fond of oysters, tomatoes, or olives; they began by using a small amount and then gradually educated their sense of taste, until a fondness for them was acquired. Habit has much to do with this; we like and dislike those things which we are in the habit of seeing other members of the family like and dislike. Disease often perverts the taste, so that persons will drink vinegar, and eat chalk and plaster.

Impressions of Taste Remain. If a very sweet or a very bitter substance be placed in the mouth and then removed from it, the taste is retained for some time. Therefore, if one substance be tasted and then quickly followed by others of different tastes the impressions will be confused. If the taste of the first was well marked it may impart its qualities to those following. Therefore, to take a medicine which has a disagreeable odor and taste, first take into the mouth some strongly flavored substance; hold the nose, to shut out the odor, and then swallow the dose. In this manner, for reasons already given, there will be little recognition of the drug. Young persons often form the habit of eating cloves and other spices: this is very harmful, not only because it is likely to injure the sense of taste, but also because it seriously disturbs the action of the stomach.

Tobacco and Taste. Tobacco blunts the sense of taste. This is exactly what we should expect, for the papillæ of the tongue become saturated with the tobacco flavor, and the taste buds are impaired by their contact with the poisonous properties of the nicotine. The taste of tobacco is continuously in the mouth, and in order to taste other substances, they must be highly spiced; this leads to disorders of the stomach as already described.

The Nose. The sense of smell is located in the nose. The two openings which lead into the nose are called the nostrils. They are sur rounded by a number of fine hairs, which aid in keeping foreign bodies from entering the nasal cavities. The framework of the nose con- sists of bone and cartilage. A thin wall, called the septum, divides the interior into two cavities; these are irregular in shape and extend from the nostrils, in front, to the upper part of the pharynx, behind.

FIG. 108. Transverse section of the framework of the nose : (1) the nasal cavities. On the outside of each cavity are the curved turbinated bones ; (2) the bones forming the roof of the mouth and the floor of the nasal cavities. The black represents the bone, the lighter shade represents the mucous membrane covering the bone.

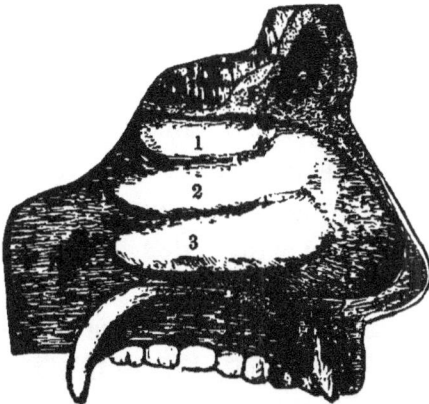

FIG. 109. The outside of the left nasal cavity, showing the three turbinated bone.

The inside of each, or the side toward the median line, is smooth, because the septum itself is smooth ; but the outside is most irregular, owing to the presence of three curved or scroll-like bones, called the turbinated bones. These are well shown, as viewed from the side, in Fig. 109. Lining each nasal cavity is a mucous membrane, which is especially thick over the turbinated bones.

The Olfactory Nerve. The nerve of smell is called the olfactory nerve. It arises from the lower and front part of the brain, and passes down through minute openings in the skull just beneath it. Its fine fibers are distributed to the upper half of the mucous lining of the nose; therefore only this part of the nasal cavity has the sense of smell. The lower half is endowed with ordinary sensibility so that, when irritated in any way, it causes the reflex act of sneezing.

The Sense of Smell. We know little about the action of odorous bodies; we cannot see an odor, neither can it be measured. Musk has been placed in an open dish in a room for many months, during which time it was constantly giving off a powerful odor, yet its weight was not diminished during the entire time.

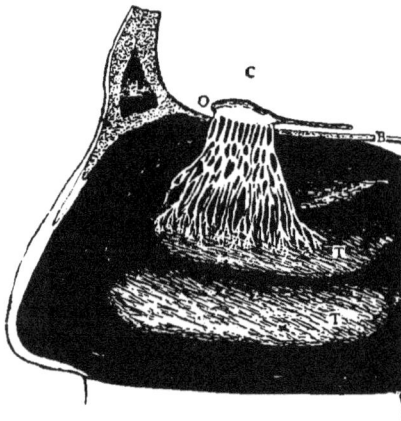

FIG. 110. The right nasal cavity, showing the termination of the olfactory nerve: T, the turbinated bones, as represented in Fig. 109; o, the olfactory bulb, lying beneath the front part of the cerebrum, c; B, the bony floor on which rests the cerebrum.

We simply know that certain substances give off a gaseous or vaporous material; that this material pervades the surrounding atmosphere; and that it is often inhaled with the inspired air. We know further, that it may produce a peculiar effect on the olfactory nerve, giving us the sense of smell.

Conditions Affecting this Sense. In order to appreciate the odor of a substance, it must be brought to the olfactory nerve in a gaseous or vaporous condition. Solid or liquid bodies, in the nose, do not produce any sense of smell. This is easily proved by filling the nose with rose water; after so doing, no odor of the rose is perceived.

The continued influence of an odor blunts the acuteness of smell. This is illustrated in every-day life. Upon first entering a room we may notice the odor of escaping gas, while in a short time we become unconscious of its presence ; we notice that a room is "close" only when we first enter it. In all such cases the first impressions should be the guide. Some persons are extremely susceptible to odors of all kinds ; they not only detect the least pleasant or unpleasant odors, but they are often made ill by them. In some people the inhalation of certain powders excites violent inflammation of the nasal passages.

The sense of smell may be greatly developed. It is related of a certain boy named James Mitchell, who was born deaf, dumb, and blind, that he could accurately identify many objects, simply by the sense of smell. Repeated and short contacts of an odor with the nasal mucous membrane favor the development of the sensation. It appears that the odor impresses the olfactory nerve at the first moment of contact. Therefore repeated sniffings, or rapid inspirations through the nose, give a more exact impression of the odor. If substances of different odors be placed near the nostrils, they will not be confused, but first one odor and then the other will be perceived.

Use of Sense of Smell. The sense of smell is of use in many ways : it aids in the choice of foods, for, as a rule, food which has a tainted odor is unfit for use ; and it aids in the detection of impurities in the air. It is placed at the very entrance of the air into the body to give warning of approaching danger. It is true that this sense does not warn us of the poisonous agents in the air which cause the contagious diseases, yet it does give notice of offensive vapors which are dangerous to inhale. Vapors which are irritating to the nose would be much more so to the more delicate tissues of the lungs. It is safe to say that in the majority of cases, disagreeable odors mean dangerous odors.

Often Highly Developed. In some of the lower animals the sense of smell is developed to a marvelous degree. The capabilities of the dog are none the less wonderful because so commonly observed. Nearly every one can tell some wonderful stories of this animal, where the power to detect obscure odors is concerned. We all know how he will return home after having been taken away long distances ; how he will follow many feet behind his master through crowded streets and into crowded halls ; how he will recognize clothing ; and how he will follow the trail of the fox for many miles. Other animals, as the lion, can tell of the approach of man, or if any prey be near ; while the deer can detect " in the air " the approach of an enemy when a great distance away.

Catarrh. Catarrh is usually the result of repeated inflammation of the mucous membrane of the nose. As a result the tissues are swollen and thickened, so that it is very difficult to breath through the nose. This makes it necessary to breathe through the mouth, which is a great source of throat trouble. Catarrh also results from deformed bones of the nose. The septum is bent over to one side, nearly or completely closing that nasal cavity. This may necessitate breathing through the mouth. The great majority of cases of catarrh can be cured by proper treatment.

QUESTIONS.

1. Give a general description of the tongue.
2. How many varieties of papillæ on it?
3. What are found in these papillæ?
4. Where are the taste buds found? What are they?
5. What varieties or qualities in the sensation of taste?
6. Why cannot dry substances be tasted?
7. What fluids aid in dissolving the substances to be tasted?
8. Illustrate how the sensations of taste and smell are often confused.
9. Do impressions of taste remain? Illustrate.
10. How does tobacco affect the taste?
11. Where are the turbinated bones ?
12. What is the name of the nerve of smell?
13. What part of the nasal cavity is endowed with the sense of smell ?
14. Is much known about the action of odorous bodies?
15. State some of the things known.
16. Give some of the conditions affecting the sense of smell.
17. How is this sense useful to us?
18. Give some illustrations showing how highly it may be developed.

CHAPTER XXXI.

THE SENSE OF HEARING.

The Organ of Hearing. The organ of hearing consists of three parts : the external ear ; the middle ear ; and the internal ear. The vibrations of the air are collected by the external ear, received by the middle ear, and transmitted through its bones to the inner ear. The inner ear contains the termination of the nerve of hearing, or the auditory nerve.

The External Ear. The external ear consists of a framework of cartilage which is loosely attached to the bones of the head and to the auditory canal. The ear can be slightly moved by the action of certain muscles, although in man this is barely perceptible. In the lower animals the movements are very extensive. The ear is quickly changed from one position to another to better catch the sound coming from any quarter. In these animals the ears aid greatly in giving expression.

The auditory canal is about an inch or an inch and a quarter in length, and extends from the external opening to the middle ear. Near the orifice are a number of fine hairs, and farther in are the openings of glands which secrete the earwax. Both the hairs and wax are for the protection of the ear, keeping out small insects, dust, and other foreign bodies.

The Middle Ear. The middle ear, or tympanum, is an irregular shaped cavity about one half an inch in length, and one fourth an inch from side to side. It is called " the drum of the ear," because it contains air and has a thin membrane over one part of it which is easily affected by wave sounds. The middle ear, or the drum, is separated from the auditory canal by a thin membrane,

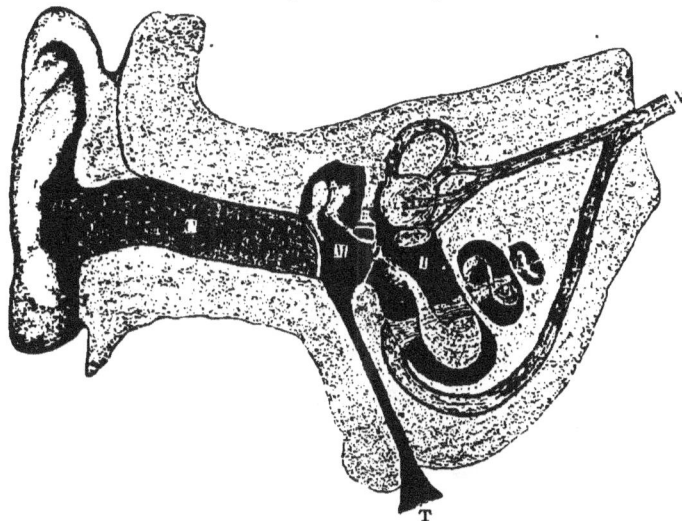

FIG. 111. The ear : c, the auditory canal, that leads to the middle ear ; m, the middle ear, or drum. The tympanic membrane is the curved white line to the left of the letter m; i, the inner ear; n, the auditory nerve going to the brain; t, the Eustachian tube, leading from the middle ear to the upper part of the pharynx.

called the tympanic membrane ; this is often called the " drum," but incorrectly so, as it is only the thin membrane over the head of the drum ; it is elastic, and so thin that it is nearly transparent. A study of Fig. 111 will aid in locating the parts already mentioned. The external ear with the auditory canal, C, is very evident. The middle ear, M, is separated from the outer ear, by a curved white line which represents the tympanic

membrane. Directly above the letter M, are three minute bones described below; the one at the left resembles
a hammer; the middle one, an anvil; and the one at the
right, a stirrup. A tube extends from the middle ear to
the throat. At the right of the middle ear is the inner
ear, at I ; it is most complicated in its structure, and is
separated from the middle ear by a thin membrane
against which the stirrup bone rests.

The tympanic membrane is often diseased from inflammation of the middle ear. Not infrequently it has
minute openings through it, while sometimes it is nearly
all destroyed. It is the function of the tympanic membrane to catch the sounds entering the external ear.
As they strike the membrane, they cause it to vibrate,
and these excite a corresponding vibration in the parts
beyond.

Bones of the Middle Ear. In the middle ear are three
bones, so minute that all together they weigh but a few
grains. Yet they
give attachment to
minute muscles,
have movable joints,
and perform most
important work.
They very much re-
semble three well
known articles;
hence they are called
the malleus, or ham-

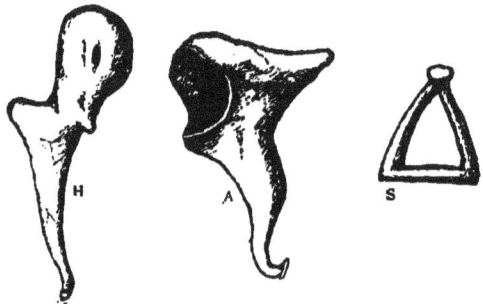

Fig. 112. The three bones of the middle
ear: H, the hammer, or malleus; A, the anvil, or incus; S, the stirrup, or stapes.

mer; the incus, or anvil; and the stapes, or stirrup. The
tympanic membrane is attached to the handle of the
hammer; the hammer, to the anvil; and the anvil, to

the stirrup ; thus a chain of bones is established from the tympanic membrane across the cavity of the middle ear. The outer end of this bony chain is attached to a membrane, and so is the inner end. Beneath the inner membrane, just opposite the stirrup, in the inner ear, is

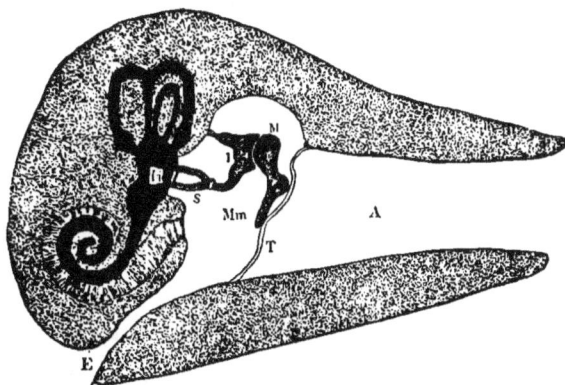

FIG. 113. The middle and inner ears, from a different view and on a larger scale than Fig. 111. A, the auditory canal of the outer ear; T, the tympanic membrane; M m, the middle ear; E, the Eustachian tube; I i, the inner ear, surrounded by bone; s, the stapes; I, the incus; M, the malleus.

a fluid. Therefore the vibrations of the air at last come to affect this fluid in the inner ear. A reference to Fig. 113 will make this clear ; the sound enters the external·ear and passes down the auditory canal, A, and strikes against the tympanic membrane, T, throwing it into vibrations ; these vibrations are communicated to the handle of the hammer, M ; thence to the anvil, I ; thence to the stirrup, S ; and thence to the membrane to which the stirrup is attached. As this membrane vibrates it throws a fluid beyond it into corresponding vibrations ; and these so affect the endings of the auditory nerve that we have at last the sensation of hearing.

The Eustachian Tube. The middle ear is not a closed cavity. It communicates with the pharynx by means of a passage, called the Eustachian tube. The tube and the middle ear are lined with a mucous membrane, and the former opens into the upper part and on the side of the pharynx directly behind the opening of the lower part of each nasal cavity. The object of this tube is to keep the air in the middle ear of the same density as that outside the ear. In the healthy ear, therefore, the air in the external ear and that in the middle ear are of the same density, with only a thin vibrating membrane between.

The Eustachian tube is ordinarily closed, opening only during the act of swallowing, and thus allowing the passage of air to the middle ear. It is often closed, as a result of chronic catarrhal affections, so that it does not open for the passage of air through it; this condition is one of the causes of deafness; but it is often possible to relieve such cases by the use of instruments with which air may be forced through the tube into the middle ear.

The Internal Ear. The internal ear is the most essential part of the organ of hearing, for it contains the terminal fibers of the auditory nerve. Its parts are deeply seated in the bones of the skull and are most intricate and complicated in structure.

The Sense of Equilibrium. A certain portion of the inner ear consists of three bony tubes, called the semicircular canal. When these canals are injured in the lower animals it is found that the animal rolls its head from side to side, up and down, while all its movements are irregular. It appears to be unable to direct its movements; for a bird thus injured will experience great difficulty in walking to the food placed near, and in picking it up. The animal sees well and appears to hear well, but it reels and falls, acting as if it were dizzy. From these observations and from some made on man, it is believed that the semi-circular canals contain nerves which enable the body to maintain its proper poise, or balance, a condition just opposite that of being dizzy. It is probable that this function residing in the middle ear is connected with the similar function of the cerebellum.

Care of the Ears. One of the most common causes of injury to the ears is the introduction of pins, or other hard substances, into the auditory canal for removing the earwax. The ordinary washing and wiping with a towel is sufficient to insure perfect cleanliness, while the constant introduction of hard substances is likely to set up an irritation which may eventually impair the hearing. Currents of cold air blowing into the ears may do great harm by chilling the sensitive tympanic membrane. The contact of cold water with this membrane often causes the earache, or acute inflammation of the middle ear; for this reason it is always better to place pieces of cotton in the ears before diving, or bathing in the surf.

Foreign bodies in the ear are not always easily removed. Insects are best removed by having warm sweet oil gently poured into the auditory canal; this will either drive the insect out or kill it; probably both. Physicians should always be consulted for the removal of other foreign bodies. The firing of a cannon has caused deafness to those standing near. Boxing the ears is especially dangerous ; the sudden forcing the air into the ear may rupture the tympanic membrane and seriously impair the hearing.

The value of acute hearing to the child cannot be over-estimated ; it bears a more important relation to the intellectual development than the eye itself. This is proved by the fact that children born blind often ex-hibit remarkable faculties of the mind ; while generally the mental deficiencies of those born without hearing are very marked. The hearing can be greatly developed by training. The acute ear of the uneducated Indian will detect the approach of moving bodies long before the untrained ear is made aware of their approach ; and the skilled musician can detect the most delicate variation from the proper sound of a note.

QUESTIONS.

1. Name the parts composing the organ of hearing.
2. Describe the auditory canal.
3. Why is the middle ear called the drum of the ear?
4. Where is the tympanic membrane located?
5. Give the names of the bones of the middle ear.
6. Describe the effects produced by sound entering the external ear.
7. What is the object of the Eustachian tube?
8. Its closure often causes what ?
9. Where is the internal ear situated ?
10. What is the function of the semicircular canals?

CHAPTER XXXII.

THE SENSES OF TOUCH: TEMPERATURE: WEIGHT: PRESSURE: COMMON SENSATION: AND PAIN.

The Sense of Touch. A reference to the chapter on the skin will recall the fact that in some of the papillæ of the skin are nerve fibers; in Fig. 92, at 4, is such a nerve termination. These papillæ are called the touch corpuscles, as they are especially concerned in the sense of touch. It must be remembered, as shown in Fig. 92, that there is only a thin layer of cells between these corpuscles and the surface of the skin; therefore it is readily understood that any impression made on the outer surface of the skin is almost in direct contact with one of these touch corpuscles.

The touch corpuscles are very numerous on the palms of the hands and palmar surfaces of the fingers, while they are the least numerous on the back. The sense of touch seems most acute on the tip of the tongue. It is the least complicated of any of the senses and is the one first developed in the child. It is in constant use, bringing us into the closest relations with external objects. By its use, we learn the size, figure, solidity, and smoothness, as well as many other qualities of bodies.

The sense of touch is capable of being highly developed, especially if great reliance has to be placed upon it; it is developed to the greatest degree in the blind. They soon learn to read by passing the fingers over raised letters; to recognize persons by feeling their faces; to distinguish plants by touching them; and to become expert musicians. The blind sculptor Gonelli is said to have modeled beautifully, relying on the sense of touch alone.

This sense is very accurate, and less liable to deceive than the other senses. The eye and ear often convey to us vague and many times wrong impressions; but the touch at once corrects these and gives us the true condition. This is well illustrated in cases where persons who were born blind have had their sight restored by some surgical operation. They could not tell a globe from a round card, nor a cube from a square, nor a disk from a circle; but as soon as the hands were placed on these bodies the errors were immediately discovered.

Degrees of Acuteness of Touch. Those parts are most sensitive to touch which have the greatest number of touch corpuscles. The degrees of delicacy are measured by means of a pair of compasses with blunted points. The two points are touched at the same moment to the skin, while the eyes are closed; they are gradually brought nearer together until both points are felt as one, when their distance apart is noted. The same experiment can be performed, though less accurately, with pins. Two pins are held with their points at least an inch apart, and then pressed lightly against the skin on the back of the wrist of another person. Repeat the experiment, bringing the points nearer together each time. Soon the person will declare that he feels but one point instead of two. The distance between the points will give the degree of acuteness of touch for that part of the body. In this way it has been proved that the shortest distance at which the two points of a compass can be distinguished as double is as follows: on the tip of the tongue $\frac{1}{25}$ of an inch; on the inside of the tips of the fingers $\frac{1}{5}$ of an inch; on the palm of the hand about $\frac{1}{3}$ of an inch; on the cheek about $\frac{3}{4}$ of an inch; and on the back over 2 inches.

Touch in the Lower Animals. All forms of animal life appear to have this sense developed to a greater or less degree. In the cat there are special organs endowed with unusual sensitiveness; these are called the whiskers. The long hair of the cat and other animals renders the general surface of the skin little adapted for the sense of touch; hence, they have been provided with a few long hairs, called feelers, which are in close connection with nerve fibers in the skin. The touch is extremely sensitive in the trunk of the elephant; while many small insects have special organs, or feelers, for this sense.

The Sense of Temperature. The temperature sense makes us acquainted with all the variations in the temperature of the skin. The skin has a proper temperature of its own and when this is maintained we are unconscious of either heat or cold; but if any part of the skin rises above its proper temperature we feel warm, and if it falls below it we feel cold. When a body is applied to the skin which takes heat from it a sensation of cold is produced; while if a body imparts heat, warmth is experienced. The sense of temperature appears to be principally in the skin, the mouth, the throat, and at the entrance of the nose. The appreciation of temperature varies for the different parts of the body, as for instance, hot applications which would be intolerable on the face can be borne when applied to the scalp.

The sensations of heat and cold are sometimes strangely confused. If the hand be dipped in very cold water, and then dipped again in water a number of degrees warmer, there is first a feeling of warmth and then of cold; if one finger be dipped in warm water, the feeling of warmth will not be nearly so great as it would if the whole hand were immersed; if two equal weights be lifted in the hands, one warm and the other cold, the latter will be declared to be heavier.

The Sense of Weight.   This is also known as the muscular sense. It informs us of the amount of muscular contraction necessary to lift a body; while it depends partly on the sense of pressure and partly on common sensibility. It is, therefore, about midway between a special sense and the common sensation of the body. By placing weights in each hand and then raising and lowering the hands, one becomes conscious of a certain amount of resistance. The muscular exertion required to lift the body gives us the sense of weight; and by practice it is possible to distinguish very slight differences in the weights of bodies, even of those as light as coins.

The Sense of Pressure.   The sense of pressure enables us to judge of the amount of weight or pressure, on different parts of the skin. To illustrate this, the hand, or the part being tested, must rest on the table, or must be supported in some way. The various parts of the body differ exceedingly as to the amount of weight required to make itself felt: the most acute portion is the forehead ; next, the temples; then the back of the head ; and lastly, the forearm.

Common Sensation. The term " common sensation " refers to all parts of the body, which have sensitive nerves that are capable of causing pleasant or unpleasant sensations. These cannot be compared to the special senses already described ; in fact, they are difficult to describe at all. We have many of these common sensations, each one of a character peculiarly its own ; thus we speak of the sensations of hunger, of thirst, of pain, of fright, of fatigue, of illness, and of health.

Pain. If any of the nerves of sensation be disturbed it produces a sensation called pain. If a sensitive nerve be cut in any part of its course it produces pain which is always referred to the place where the nerve ends, rather than to the point injured. Thus, touching the ulnar nerve, the " crazy bone " at the elbow, causes a pain in the little finger and part of the adjoining finger ; these are the parts in which the nerve terminates. After amputation of limbs, it often occurs that one of the severed nerves gives rise to pain ; in this case the afflicted person says he feels the pain in the amputated fingers or toes. It is not known what causes the varieties of pain ; for some are sharp and cutting, while others are dull and throbbing.

Dreadful as it is, still pain is a valuable bodyguard. It tells of the approach of danger, and points to disease when it is present. It may be stated that any so-called stimulus will cause pain if applied beyond the normal limit and to an excessive degree. Light is the stimulus for vision ; yet strong light, as the glare of the sun, is at once painful. Sounds falling on the ear may awake the most pleasant memories; yet loud and long continued sounds soon become positively painful. The ordinary contraction of a muscle is free from pain; yet the rapid and violent muscular contractions in spasms are intensely painful.

The nerves of sensation are the great protectors of the body. Without them and their accompanying pain we should cut, burn, bruise, and otherwise disfigure the body in many ways as we go about our daily work. Pain keeps us from pursuing many harmful courses, and thus aids greatly in protecting the body. It is an unaccountable fact that some persons suffer more pain than others who have the same diseases or injuries; so it follows that the severity of the suffering does not always indicate a corresponding intensity of the disease or injury.

Animals appear to suffer pain, especially those animals which are most noted for their intelligence, as the dog and the horse. The finer bred the animal, so much the more susceptible is it to pain. The thoroughbred and blooded horse appears to suffer great pain, while the ordinary work-horse may be perfectly indifferent. But those creatures low in the scale of animal life do not exhibit evidence that they suffer much pain.

CHAPTER XXXIII.

ADDITIONAL TESTIMONY AGAINST ALCOHOL AND TOBACCO.

From the Medical Profession. In Canada the medical profession is awake to the growing evils of intemperance. In the city of Montreal, twenty-five professors in medical colleges, and over seventy physicians of the city, united in the following declaration against alcohol : —

"We, the undersigned members of the medical profession in Montreal, are of opinion —

"1. That a large proportion of human misery, poverty, disease, and crime is produced by the use of alcoholic liquors as a beverage.

"2. That total abstinence from intoxicating liquors, whether fermented or distilled, is consistent with, and conducive to, the highest degree of physical and mental health and vigor.

"3. That abstinence from intoxicating liquors would greatly promote the health, morality, and happiness of the people."

In the cities of New York and Brooklyn, many of the most prominent and influential members of the medical profession unhesitatingly declare against the use of alcoholic beverages. Notice the strong language in the following declaration : —

" 1. In view of the alarming prevalence and ill effects of intemperance, with which none are so familiar as members of
the medical profession, and which have called forth from eminent English physicians the voice of warning to the people of
Great Britain concerning the use of alcoholic beverages, we,
the undersigned members of the medical profession of New
York and vicinity, unite in the declaration that we believe
alcohol should be classed with other powerful drugs; that when
prescribed medicinally it should be with conscientious caution
and a sense of grave responsibility.

" 2. We are of opinion that the use of alcoholic liquor as a
beverage is productive of a large amount of physical disease;
that it entails diseased appetites upon offspring; and that it is
the cause of a large percentage of the crime and pauperism of
our cities and country.

" 3. We would welcome any judicious and effective legislation,
State and national, which should seek to confine the traffic in
alcohol to the legitimate purposes of medical and other sciences,
art, and mechanism."

This paper was signed by over one hundred of the
leading physicians of New York and Brooklyn. Among
the names we find many that are familiar to the medical
profession of the whole world, and some that are known
wherever the subject of public health is discussed. A
large number of the signers is composed of men who
have taught five years in medical colleges, practised in
large hospitals, and been actively engaged as members
of Boards of Charity and Public Health Associations. As
teachers they have become acquainted with the scientific
side of the question, and as practitioners they have become familiar with the practical results of the use of
strong drink. No men are better qualified to judge of
the evils of strong drink than those of the medical pro-

fession, and the declarations of these men is against the use of alcoholic liquors as a beverage.

Alcohol is a Poison. Dr. E. W. Lambert, the medical director of the Equitable Life Insurance Company, writes to a prominent member of the United States Senate as follows: —

" To speak chemically, alcohol is a concentrated hydrocarbon, and needs a great deal of physical labor to dispose of it in the animal economy. I have noticed that men who are given to the daily use of alcohol degenerate faster than those who abstain from its use. They are more liable not only to chronic degenerations (such as fatty livers, fatty kidneys, and the like), but they are also more liable to be attacked by acute diseases ; and acute diseases are much more likely to prove fatal to the users of alcohol than to those who do not use it.

"Take for illustration a young friend of mine, who commenced the use of alcohol about the age of twenty-one years. He died after two days' illness. When I came to examine his body after his death, I found that all his internal organs belonged to a man of the age of seventy years, and not to a man of forty, the age at which he died. I have noticed that steady users of alcohol are very much more apt to die between the ages of forty and fifty years of some acute disease than those who do not use it as a beverage.

"I have said nothing as yet concerning the danger which every one undergoes who uses alcohol regularly. . . . If he once gets the appetite, there is nothing on the earth, or above it, or under it, that he will not do in order to gratify this morbid appetite. He will lie or steal, or see his family go to ruin with perfect equanimity, provided he can satisfy this inordinate craving for alcohol."

Dr. Dods, of England, said before a committee of the House of Commons: —

"Writers on Medical Jurisprudence rank alcohol among narcotico-acrid poisons, of which small quantities, if repeated, always prove more or less injurious, and agree that the morbid appearances seen after death occasioned by ardent spirits exactly agrees with those that result from poisoning caused by any other substance of the same class."

Medical men are familiar with the physiologies of the renowned Dr. W. B. Carpenter, of London. His works on mental physiology have made him rank high as a philosopher; while his exhaustive treatises on microscopy illustrate how closely he observes even the minutest things. Dr. Carpenter concludes his observations on the subject of alcohol as follows: —

"The introduction of alcohol into healthy blood can do nothing but mischief; no one who is familiar with the action of poisons upon the living animal body, and has made the nature of that action a special study, has the smallest hesitation in saying that *alcohol is a poison*."

Alcohol is of no Use. Dr. William Pepper, of the University of Pennsylvania, is another author of great prominence. His exhaustive work on the Practice of Medicine has made for him an enduring name as a profound scholar and keen observer. He testifies as follows: —

"One of the worst features of the action of alcohol in a large majority of young persons is that, though taken in small amount, and even in the form of light wines or beer, its first agreeable effect is followed by a feeling of lassitude and depression, readily mistaken for debility, and suggesting a repetition of the stimulant. But these unpleasant feelings are the direct result of the presence in the blood and tissues of poisonous matters, coming from the imperfect digestion of the alcohol, or of food with whose complete assimilation the dose of alcohol has interfered. The habitual use of alcoholic beverages by healthy persons is highly injurious, and involves the risk of developing serious disease. I am, indeed, satisfied that all persons in good health are *better without alcohol*, in any form or in any amount, as a beverage."

From the hundreds of opinions of individual medical men we select the following as illustrating the fact that scholarship and experience alike condemn the use of strong drink. Dr. N. S. Davis, of Chicago, takes the most advanced ground, going so far as to state that alcohol is not even desirable as a medicine. Dr. Davis is not only most skilled in his profession, but is also a man of great general learning. He has been honored with the highest gifts in the power of the profession to bestow upon him, while his opinion is acknowledged by all to be of the greatest weight. To what does this eminent scholar testify? —

"I have been constantly engaged in the practice of medicine a little more than *fifty years*, embracing both private and public hospital practice, and have demonstrated by the last forty years of actual experience that no form of alcoholic drink, either fermented or distilled, is necessary or desirable for internal use in either health or in any of the varied forms of disease; but that health can be better preserved and disease be more successfully treated without any use of such drinks. And while it is true that during the last thirty years I have not prescribed for internal use the aggregate amount of one quart of any kind of fermented or distilled drinks, either in private or hospital practice, yet I have continued to have abundant opportunities for observing the effects of these agents as given by others with whom I have been in council; and simple truth compels me to say that I have never yet seen a case in which the use of alcoholic drinks either increased the force of the heart's action or strengthened the patient beyond the first thirty minutes after it was swallowed."

Alcohol Shortens Life. The President of the Connecticut Mutual Life Insurance Company has given the following testimony : —

"Among the persons selected with care for physical soundness and sobriety, and who are, as a rule, respectable and useful members of society, the death-rate is more profoundly affected by the use of intoxicating drinks than from any other one cause, apart from hereditary.

"I protest against the notion so prevalent and so industriously urged that beer is harmless, and a desirable substitute for the more concentrated liquors. Its use is an evil only less than the use of whiskey, if less on the whole, and its effect is only longer delayed; its incidents not so repulsive, but destructive in the end. In one of our largest cities, containing a great population of beer drinkers, I had occasion to note the deaths among a large group of persons whose habits, in their own eyes and in those of their friends and physicians, were temperate, but they were habitual users of beer. When the observation began they were, upon the average, something under middle age, and they were, of course, selected lives. For two or three years there was nothing very remarkable to be noted among this group. Presently death began to strike it; and, until it had dwindled to a fraction of its original proportions, the mortality in it was astounding in extent, and still more remarkable in the manifest identity of cause and mode. There was no mistaking it; the history was almost invariable, — robust, apparent health, full muscles, a fair outside, increasing weight, florid faces; then a touch of cold, or a sniff of malaria, and instantly some acute disease, with almost in-

variably typhoid symptoms, was in violent action, and ten days
or less ended it. It was as if the system had been kept fair
outside, while within it was eaten to a shell; and at the first
touch of disease there was utter collapse, — every fiber was
poisoned and weak. And this, in its main features, varying,
of course, in degree, has been my observation of beer drinking
everywhere. It is peculiarly deceptive at first; it is thoroughly
destructive at the last."

Henry Ward Beecher said : —

"Every year I live increases my conviction that the use of
intoxicating drinks is a greater destroying force to life and virtue than all other physical evils combined."

Alcohol causes Pauperism. We little realize the great
amount of pauperism that exists in our own country.
Even to a less degree do we realize that this great
burden of society is so largely caused by strong drink.
We have some positive testimony on this point. The
chairman of the Board of Health of the State of Massachusetts sent out the following two inquiries to the town
and city authorities of the State : —

"1. What proportion of the inmates of your almshouses
are there in consequence of the deleterious use of intoxicating liquors?
"2. What proportion of the children are there in consequence of the drunkenness of parents?

He received two hundred and eighty-two replies.
Among these is the following from the superintendent
of Deer Island almshouse and hospital: "I would answer the above by saying, To the best of my knowledge,
ninety per cent, — to both questions."

The authorities of the city of Springfield reported: " We have fed 8,052 tramps.   Seldom found one not reduced to that condition by intemperance."

In the county of Suffolk, mainly the city of Boston, eighty per cent of the pauperism was caused by intemperance.   Yet we must remember that the excellency of the government, and the superiority of the schools of Massachusetts, should make pauperism more rare than in less favored States.   But even here the conclusion is forced upon us that a very large per cent of all cases of pauperism is attributable to the vice of intemperance.

Alcohol causes Crime. The United States Commissioner of Education says that " from eighty to ninety per cent of our criminals connect their course of crime with intemperance."

The Board of Public Charities of Pennsylvania said: " The most prolific source of disease, poverty, and crime is intemperance."

The Citizens' Association of Pennsylvania states that " it will not be doubted that two thirds of the pauperism and crime are justly attributed to intemperance."

The inspectors of Massachusetts State Prison testified that in 1868 " about four fifths of the number committed the crimes for which they were sentenced, either directly or indirectly, by the use of intoxicating drinks."

Judge Noah Davis, Ex-Chief Justice of New York, says that " ninety per cent of the criminal business of the courts is caused by the liquor traffic."

The Hon. A. G. Fairbanks, of Manchester, N. H., testifies that he is familiar with over one thousand cases of persons confined in jail for various offences, and he gives it as his opinion that, " directly and indirectly, it is safe to say that seventy-five per cent of all crimes can be traced to the use of alcohol as a beverage."

The Hon. William J. Mullen, of Pennsylvania, says:

"Of the half million persons who had been committed to the county prison of Philadelphia during the last twenty years, there had been about five hundred for murder; seven hundred for attempts to murder; over forty thousand for assault and battery, and over two hundred thousand for drunkenness. In nearly every case of murder or attempt to murder the parties were intoxicated."

An evidence of the bad effects of alcoholic liquors may be seen in the fact that there have been thirty-four murders within the city of Philadelphia during one year, each one of which was traceable to intemperance, and one hundred and twenty-one assaults for murder proceeding from the same cause. Of over thirty-eight thousand arrests in Philadelphia within one year, seventy-five per cent were caused by intemperance. Of 18,305 persons committed to prison within the year, more than two thirds were the consequence of intemperance.

Judge Allison says: —

"In our criminal courts we can trace four fifths of the crimes that are committed to the influence of rum. There is not one case in twenty where a man is tried for his life in which rum is not the direct or indirect cause of the murder."

Alcohol causes Wrong Expenditure of Money. In speaking of the large amount of money expended by the people of England year by year for intoxicating drinks, Cardinal Manning said: —

"Can there be a more complete waste? Expend it in the drainage of England and the culture of the land, and there

22

would be bread for the hungry mouths of the people; in the manufacture of cloth, and there would be no man and no child without a coat on his back; in the building of houses fit for human habitation, and there would not be a working man and his family without a roof over his head. Nay, I will go further. It is not only a waste, it has a harvest. It is a great sowing broadcast; and what springs from the furrows? Deaths, mortality in every form, disease of every kind, crime of every die, madness of every intensity, misery beyond the imagination to conceive."

Alcohol neutralizes Educational Agencies. The New York "Tribune," in an editorial, thus speaks of the liquor traffic : —

"It is impossible to examine any subject connected with the progress, the civilization, the physical well-being, the religious condition of the masses, without encountering this monstrous evil. It lies at the center of all political and social mischief; it paralyzes energies in every direction; it neutralizes educational agencies; it silences the voice of religion; it baffles penal reform; it obstructs political reform. . . . There is needed something of that sacred fire which kindled into inextinguishable heat the zeal of the abolitionists, and which compelled the abandonment of human slavery, to rouse the national indignation and abhorrence against this very much greater evil."

Alcohol is opposed to Good Order. In giving a history of the army of the Potomac, the writer says : —

"I had occasion to observe a remarkable difference in the appearance of the different regiments. In some cases I have found their men dirty, their camp disorderly, and their whole appearance shabby; in others, everything neat and tidy, orderly, and well-disposed. On inquiry, I have found that the difference was owing in great degree to the course the commanding officers have pursued in relation to the use of intoxicating drinks. Where, as in a great many instances, the colonel has enacted a 'prohibitory law,' and forbidden the admission of liquor into the camp, I find everything in the best condition, the best health, the best order. I was much gratified to find that a great many officers and soldiers abstained entirely, — not because they were compelled, but because they chose to do so. No small number of officers in high command are teetotalers. The result of my observations in regard to temperance in this great army at Washington is, that the common-sense of both officers and men is strongly in favor of prohibition; and wherever it has been enforced with fidelity and vigilance, it has been in the highest degree beneficial."

The Conclusion of the Whole Matter. For many years it was the privilege of the author to be associated with the eminent Dr. Alonzo B. Palmer, who for so many years was dean of the medical department of the University of Michigan. Dr. Palmer was like Dr. Davis, in that he was a man of wide observation and scholarly attainments. Shortly before his death he wrote as follows:

"If chloroform is a narcotic, alcohol is a narcotic; if chloroform is an anæsthetic, alcohol is an anæsthetic; if one is essentially a depressing agent, so is the other. Their strong resemblance no one can question. The chief difference is, that the alcoholic narcosis is longer continued, and its secondary effects are more severe.

"There is a connection, often marked, in the use of the different narcotics. The alcohol habit tends to produce the opium habit, and the reverse; one may be substituted for the other, and the two are often indulged together. The same principle, to a greater or less extent, applies to the widespread tobacco habit, and to the less prevalent chloroform, chloral, and hasheesh habits. The indulgence in any one begets a tendency to indulge in others. The habitual use of any of them produces a constitutional narcotic state, different from the normal.

" We thought, and we may sometimes still think, it makes us witty. We know from observation it makes men silly.

" We thought it brightened the intellect and might make men wiser. We find that in the long run, at least, it dulls the intellect and makes men foolish.

" Wine has been called the 'milk of age,' and we thought it supported advanced life. We know that the aged live longer and retain their powers better without its use.

" As a medicine, or prophylactic measure, we thought it protected against epidemic diseases. We now know it invites attacks.

" We thought it prevented and even cured consumption. We know it is the most frequent cause of at least one form of that disease, — fibroid phthisis.

" We thought, moderately used, it was good for many things. Those who have given most careful attention to the subject believe it is good for very few things.

" The demonstrations of modern science have shown the truth of the ancient saying of the Wise Man. " Wine is a mocker, strong drink is raging, and whosoever is deceived thereby is not wise."

Cigarette Smoking often does Infinite Harm. We clip the following from one of the leading medical journals of the world, — the London " Lancet."

" Scarcely less injurious, in a subtle and generally unrecognized way, than the habit of taking alcoholic drinks between meals, is the growing practice of smoking cigarettes incessantly. The truth is that, perhaps owing to the way the tobacco leaf is shredded, coupled with the fact that it is brought into more direct relation with the mouth and air-passages than when it is smoked in a pipe or cigar, the effects produced on the nervous system by a free consumption of cigarettes are more marked and characteristic than those recognizable after recourse to other modes of smoking. A pulse-tracing, made after the subject has smoked a dozen cigarettes, will, as a rule, be flatter and more indicative of depression than one taken after the smoking of cigars. It is no uncommon practice for young men who smoke cigarettes habitually to consume from eight to twelve in an hour, and to keep this up for four or five hours daily. The total quantity of tobacco used may not seem large ; but, beyond question, the volume of smoke to which the breath-organs of the smoker are exposed, and the characteristics of that smoke as regards the proportion of nicotine introduced into the system, combine to place the organism very fully under the influence of the tobacco. A considerable number of cases have been brought under our notice during the last few months in which youths and young men who have not yet completed the full term of physical development have had their health seriously impaired by the practice of almost incessantly smoking cigarettes. It is well that the facts should be known, as the impression evidently prevails that any number of these little ' whiffs ' must needs be perfectly innocuous, whereas they often do infinite harm."

Tobacco affects Scholarship. Some careful observations have been made in many of our public schools and colleges concerning the effects of tobacco on scholarship. The following is taken from a recent number of a college publication : —

" Statistics as to the effects of tobacco-smoking upon students have been collected at Amherst and Yale colleges. The non-smokers at Amherst are of greater weight than the smokers; they are superior in chest-girth to the smokers, and their lung-capacity is higher. The non-smokers are more athletic than the smokers, and more successful in athletic sports. The non-smokers at Amherst, as at Yale, have also an advantage over the smokers in mental power and in scholarship. The facts recently collected in American colleges concerning the physiological and physical effects of the tobacco-smoking habit are instructive to the young men who go to college, and also to those who do not."

CHAPTER XXXIV.

EMERGENCIES.

It often happens that a knowledge of anatomy and physiology, applied in times of emergency, may save a life. To act promptly and properly is the demand. In any case of emergency obtain some knowledge of the case before sending for the physician: by so doing he will be able to bring with him the necessary appliances and remedies. With a knowledge of the following suggestions, one may aid the physician, and possibly prevent a fatal termination.

Bleeding. Bleeding from the nose is the most common and the least dangerous of all hemorrhages. It is generally sufficient to apply cold water to the forehead, and over the nose, or back of the neck, and remain quiet for a short time in the sitting posture. If this does not bring relief, then compress the nostrils for a few minutes; if the blood clots in the nose, allow the clots to remain for a few hours. Placing the hands and feet in water as hot as can be borne, will often immediately stop a severe nosebleed. If the bleeding still continues, a physician should be called, who may find it necessary to plug the nostrils with gauze or with cotton.

Bleeding after the extraction of a tooth is sometimes profuse; but, as a rule, it is easily controlled by employing pressure over the wounded gums, or by packing the cavity with cotton or gauze. A small pad is placed over the bleeding surface and the jaws closed firmly over it. The pad should be so thick that when the mouth is closed there will be firm pressure over the bleeding point.

Bleeding from the surface of the body can generally be controlled by pressure. This can be applied by the fingers, or, if the wound be over a bone, a piece of cloth may be folded into a small pad and held tightly to the wound by a bandage. If the bleeding be slight, frequent applications of cold water will probably be sufficient to check the flow.

Bleeding from an artery is recognized by the blood flowing in spurts or jets; from a vein, by its steady flow.

Burns and Scalds. Great relief is obtained in cases of burns and scalds by covering them with soft cloths saturated with a solution of common soda, — a tablespoonful of soda to a cup of water. An application of cream or a thick covering of dry flour will give relief. A liniment of equal parts of sweet oil and lime water is also very useful.

Bruises. Applications of cold water, or powdered ice in bags is beneficial. The ordinary extract of witch-hazel is largely used in such cases. Moist applications should be used only for a short time, immediately after the injury; if continued for a long time, they lower the vitality of the parts.

Convulsions. Make no attempt to hold the patient quiet, but simply prevent him from injuring himself. In nearly all cases perfect quiet is the principal thing. If the convulsions occur in children, a warm bath will often give immediate relief.

Dislocations and Fractures. If there be reason to believe that a joint has been dislocated, or a bone broken, the injured parts should be kept perfectly quiet until a physician arrives. It is better to wait a few hours for his arrival rather than to handle the parts in order to learn what is the matter.

Fainting. The principal thing to remember is that in cases of fainting there is not enough blood in the brain. Therefore, those things should be done which will promote a flow of blood to the head. Place the patient on his back and keep the head low, certainly as low as the body; do not raise the patient's head until he has fully recovered. Never give alcohol or brandy in such cases; a cup of hot coffee is far better. Dashing cold water in the face, or holding an open bottle of ammonia near the nose will perhaps aid.

Fire. If a person's clothing be seen on fire grasp the nearest rug, shawl, blanket or cloak, and wrap it tightly about his body. The flames may be smothered by rolling the person on the ground.

Fish Hook. The best way to remove a fish hook that has entered the flesh beyond the barb, is to push the point through the skin, cut the hook off below the barb, and withdraw the remainder.

Frost Bite. Rub the frozen parts with snow, or very cold water; when they begin to sting and become red then cease the rubbing, for reaction has commenced. A gradually increasing degree of warmth may now be applied.

Poisons. Whenever it is feared that poisons have been swallowed, emetics should be given at once. Give a dessert-spoonful of ground mustard mixed in a cup of warm water. If vomiting does not occur in a few min-utes, the dose may be repeated. In the place of the mustard, a tablespoonful of alum may be substituted. After the vomiting, give a glass of milk in which are the whites of two eggs well beaten.

Shock. If a person is insensible from some blow, or fall, or from fright, and yet none of the organs or tissues of the body are injured, place the patient on his back and allow a free circulation of air around him. If his head is hot, a cloth moistened with cold water may be applied to it.

Sprains. These are often of a serious nature. Bathe the parts in either hot or cold water, — whichever gives the greater relief, — until the doctor comes.

Stings. Often the "stinger" of wasps, bees, etc., re-mains in the wound. This should be removed by the fingers, or by small forceps; then apply spirits of ammonia. If no ammonia be at hand, try soda water, as recommended for burns. Boys have often found that binding on ordinary mud gives relief.

Sun Stroke. Remove the patient to a cool room or shady spot, place him on his back, slightly raise the head, and apply cold cloths to the head and face.

Wounds. In cases of wounds the principal thing to do, before the arrival of the surgeon, is to prevent too great loss of blood. Remove the clothing, if necessary, until the injured part is reached, then by pressure tem-porarily check the flow. Pressure directly over the parts or over the larger vessels leading to them is of the first importance.

INDEX.

www.ingramcontent.com/pod-product-compliance
Lightning Source LLC
Chambersburg PA
CBHW021106270326
41929CB00009B/754